SCHOOL OF AGRICULTURAL
SCIENCES AND TECHNOLOGY

Agriculture
and the
Undergraduate

Proceedings

Board on Agriculture
National Research Council

NATIONAL ACADEMY PRESS
Washington, D.C. 1992

NOTICE: The project that is the subject of this report was approved by the Governing Board of the National Research Council, whose members are drawn from the councils of the National Academy of Sciences, the National Academy of Engineering, and the Institute of Medicine. The members of the committee responsible for the report were chosen for their special competences and with regard for appropriate balance.

This report has been reviewed by a group other than the authors according to procedures approved by a Report Review Committee consisting of members of the National Academy of Sciences, the National Academy of Engineering, and the Institute of Medicine.

The National Academy of Sciences is a private, nonprofit, self-perpetuating society of distinguished scholars engaged in scientific and engineering research, dedicated to the furtherance of science and technology and to their use for the general welfare. Upon the authority of the charter granted to it by the Congress in 1863, the Academy has a mandate that requires it to advise the federal government on scientific and technical matters. Dr. Frank Press is president of the National Academy of Sciences.

The National Academy of Engineering was established in 1964, under the charter of the National Academy of Sciences, as a parallel organization of outstanding engineers. It is autonomous in its administration and in the selection of its members, sharing with the National Academy of Sciences the responsibility for advising the federal government. The National Academy of Engineering also sponsors engineering programs aimed at meeting national needs, encourages education and research, and recognizes the superior achievements of engineers. Dr. Robert M. White is president of the National Academy of Engineering.

The Institute of Medicine was established in 1970 by the National Academy of Sciences to secure the services of eminent members of appropriate professions in the examination of policy matters pertaining to the health of the public. The Institute acts under the responsibility given to the National Academy of Sciences by its congressional charter to be an adviser to the federal government and, upon its own initiative, to identify issues of medical care, research, and education. Dr. Kenneth I. Shine is president of the Institute of Medicine.

The National Research Council was organized by the National Academy of Sciences in 1916 to associate the broad community of science and technology with the Academy's purposes of furthering knowledge and advising the federal government. Functioning in accordance with general policies determined by the Academy, the Council has become the principal operating agency of both the National Academy of Sciences and the National Academy of Engineering in providing services to the government, the public, and the scientific and engineering communities. The Council is administered jointly by both Academies and the Institute of Medicine. Dr. Frank Press and Dr. Robert M. White are chairman and vice-chairman, respectively, of the National Research Council.

This project was supported by the Cooperative State Research Service of the U.S. Department of Agriculture, under Cooperative Agreement Number 90-COOP-1-5801; the U.S. Department of Agriculture Project Interact; and the Resident Instruction Section of the Division of Agriculture, National Association of State Universities and Land-Grant Colleges. Dissemination of these proceedings was supported by the W. K. Kellogg Foundation.

Any opinions, findings, conclusions, or recommendations expressed in this publication are those of the author(s) and do not necessarily reflect the view of the U.S. Department of Agriculture.

Library of Congress Catalog Card No. 92-60206
ISBN 0-309-04682-3

Copies are available for sale from:
National Academy Press
2101 Constitution Avenue
Washington, DC 20418

S523
Printed in the United States of America

Steering Committee

Karl Brandt, *Chairman,* Purdue University
C. Eugene Allen, University of Minnesota
Harry O. Kunkel, Texas A&M University
Joseph E. Kunsman, University of Wyoming
Conrad J. Weiser, Oregon State University
Paul H. Williams, University of Wisconsin

Staff

Carla Carlson, *Project Director*
Barbara J. Rice, *Associate Staff Officer*
Robert Cox, *Senior Program Assistant*
Michael K. Hayes, *Project Editor*

Board on Agriculture

Theodore L. Hullar, *Chairman,* University of California, Davis

Philip H. Abelson, American Association for the Advancement of Science

Dale E. Bauman, Cornell University

R. James Cook, Agricultural Research Service at Washington State University

Ellis B. Cowling, North Carolina State University

Robert M. Goodman, University of Wisconsin

Paul W. Johnson, Natural Resources Consultant, Decorah, Iowa

Neal A. Jorgensen, University of Wisconsin

Allen V. Kneese, Resources for the Future, Inc.

John W. Mellor, John Mellor Associates, Inc.

Donald R. Nielsen, University of California, Davis

Robert L. Thompson, Purdue University

Anne M. K. Vidaver, University of Nebraska

Conrad J. Weiser, Oregon State University

John R. Welser, The Upjohn Company

Susan E. Offutt, *Executive Director*

James E. Tavares, *Associate Executive Director*

Carla Carlson, *Director of Communications*

Barbara J. Rice, *Editor*

Preface

On many campuses, colleges of agriculture and related disciplines are undergoing programmatic changes and, more important, are reexamining the philosophy underlying their missions. They are developing a unique knowledge base that is much broader than is generally perceived—a knowledge base that is a composite of disciplines that broadly link basic sciences, natural systems, economics, business, and human resources to the more traditional production agriculture and food enterprises.

The Board on Agriculture of the National Research Council joined the U.S. Department of Agriculture (USDA) and its office of Higher Education Programs, Cooperative State Research Service, in sponsoring a landmark national conference to chart the comprehensive changes needed to meet the challenges of undergraduate professional education in agriculture. The conference, Investing in the Future: Professional Education for the Undergraduate, was held at the National Academy of Sciences in Washington, D.C., on April 15–17, 1991. Leaders from the higher education community, business, industry, and public agencies attended the conference.

After decades of association with research and extension, USDA has recently focused increased attention on the evolving educational missions of colleges of agriculture. In early 1989, under the auspices of USDA Project Interact, sponsored by the office of Higher Education Programs, a committee of 27 people holding a wide range of positions—presidents of universities, administrative heads of agriculture, department heads, faculty, and representatives from industry, USDA, and the U.S. Agency for International Development—culminated a series of studies and explored the questions concerning the requirements that will be placed on colleges of agriculture in the future. The committee, chaired by Harry O. Kunkel of Texas A&M University, reaffirmed the importance of these colleges. The committee noted that colleges of agriculture, home economics, and natural resources provide the intellectual foundations and focal points

v

for human activities related to food, agriculture, natural resources, and other life-supporting systems. This group issued a rationale and a call for a national conference.

The Board on Agriculture has previously focused its studies and activities on education—secondary education, graduate education, and doctoral and postdoctoral training in agriculture. The national symposium Future Opportunities and Challenges Unique to Science (FOCUS), held in April 1988, honored the first class of USDA national needs graduate fellows. The conference Investing in the Future: Professional Education for the Undergraduate was an extension of that pursuit. It also marked the intention of the National Research Council to place greater emphasis on science literacy and science education in the United States.

The conference was organized as a series of alternating plenary and smaller discussion sessions. A provocateur was designated to give an initial presentation during each discussion session, and the subsequent discussions and conclusions were summarized for the conference audience by a rapporteur from each session. The chapters in Part I of these proceedings comprise the presentations given at the plenary sessions. Each chapter in Part II consists of the initial presentations made at each discussion session followed by the rapporteurs' summaries. Appendix A includes short biographies of the conference participants, and Appendix B outlines the poster session that was featured during the conference.

In the agricultural, food, and environmental system, as with other segments of U.S. industry, the problems of the twenty-first century intensify more quickly than ever before, and opportunities must be seized immediately, before their peak of potential benefit has passed. The ability of the United States to resolve the spectrum of issues and related problems in agriculture—nutrition, economics and international trade, production efficiency, natural resources conservation, control of pollutants, and others—depends on depth of knowledge, the available tools and technologies, and the skill and insight to apply them.

The United States needs to invest in the future—in human capital and the scientific knowledge base—to revitalize and reinvigorate one of its leading industries, the agricultural, food, and environmental system, in its broadest sense. That objective can be met by educating all students about agriculture as well as educating others specifically for careers in agriculture.

These proceedings are a source for ideas that can contribute to the improved education of not only students of agriculture but also students throughout the higher education system.

We hope that these proceedings will thus serve to stimulate further enhancement of undergraduate professional education and continue the momentum generated by the national conference.

Karl G. Brandt, *Chair*
Conference Steering Committee

Acknowledgments

The contributions of several individuals to the conference and these proceedings warrant special mention. We acknowledge the ideas, enthusiasm, and personal time invested by numerous individuals represented by the Resident Instruction Committee on Policy. We are also grateful for the generous sharing of ideas among the 27 individuals who constituted the committee convened by the U.S. Department of Agriculture's (USDA) Project Interact to examine the requirements that will be placed on colleges of agriculture in the future.

In particular, we wish to thank Kenneth W. Reisch, associate dean emeritus, College of Agriculture at Ohio State University, for coordinating the poster presentations during the two and one-half day conference. We also acknowledge the special assistance provided by the staff members of USDA's office of Higher Education Programs, Cooperative State Research Service: Kyle Jane Coulter, deputy administrator; and Gwendolyn L. Lewis, Gail House, M. Louise Ebaugh, Stephanie K. Olson, Maxine Browne, and Anne Schumaker.

Neither the conference nor the proceedings could have attained the goals of the USDA or the Board on Agriculture of the National Research Council without the creative contributions of the speakers, the provocateurs, and the rapporteurs, who laid the foundation for discussion among conference attendees—representatives of the variety of communities related to higher education, secondary education, and the agricultural, food, and environmental system. And it will be these communities and the efforts of individuals within these communities that carry the momentum that can lead to improved education for all students.

Contents

Overview

Harry O. Kunkel

Undergraduate professional education, like education in the United States in general, has come under close scrutiny in recent years, at a time when the U.S. agricultural, food, and natural resource system faces an era of global competitiveness, inequities in worldwide food distribution, environmental and health concerns, and promising new science and technologies. These issues have also transformed colleges of agriculture from their former preoccupation with production agriculture, to a strong business approach, to greater attention to the underlying sciences. The emphasis is now on the educated person. The U.S. agricultural system, however, faces shortages of qualified scientists, engineers, managers, and specialists. Greater attention must now be given to rethinking the mission of agricultural education. This overview examines these and other issues related to undergraduate agricultural education that were stated and reiterated throughout the conference and in these proceedings.

Rethinking the Purpose of Professional Education in Agriculture

Undergraduate professional education in agriculture can offer content, context, and practice for undergraduate liberal study. It may function best by laying a foundation for understanding professional life. It may also be a model for higher education in its specified expectations of students and curriculum, its interest in improved teaching and advising, its efforts to construct an intellectual content for learning and for the application of theory and methodology, and its integration of undergraduates into a professional and disciplinary environment. Undergraduate professional education, however, is being squeezed by inadequate preparation in elementary and secondary schools, by the proliferation of master's degree studies, and by the perceived needs for general and liberal education.

The world economy is becoming ever more closely linked, and U.S. agriculture, which already is a significant player in world markets, will continue to feed and educate a developing, expanding, and needy world population. In addition to maintaining global food sufficiency, U.S. agriculture will maintain the economic competitiveness of the country and will help to move it toward energy independence. Undergraduate education in agriculture is essential to any strategy of meeting the new forces in world competition and is key to harnessing inventive genius in the marketplace. The principal international thrusts in undergraduate education today are foreign language studies, the inclusion of international content in courses, study abroad opportunities, and the implementation of area studies programs. However, progress in these areas is inconsistent. Most institutions have not determined what the optimum level of international competence should be. The typical undergraduate in colleges of agriculture is given little to aid his or her understanding of the impact of trade, global environmental impacts, and the nature of international agricultural research.

The purposes and functions of agriculture driving the changes in educational requirements will change at revolutionary speed. Agriculture will be more intensive biologically and managerially. The inputs into agriculture will follow prescriptive, rather than prophylactic, practice; biological management will replace chemical management. Agriculture will increasingly produce industrial materials and feedstocks and, as an industry, will increasingly become directed toward value-added products instead of raw commodities. Food safety and new perceptions in nutrition will result in linking production, plant and animal genetics, food processing, and transport and marketing. It is not that past undergraduate education was not good enough, but that it may not be an appropriate model to meet these changes.

The purpose, then, of the curriculum is to provide for the needs of industry in a changing world: flexibility, diversity, perspective, and values. The students needed most are those most likely to think globally, to act creatively, to value diversity, to behave responsibly, to respond flexibly, and to interact cooperatively. Open, observant, and inquisitive minds should be the goal. The graduate should have learned to visualize the whole and, in so doing, visualize what makes the whole work, which is not simply a perception of the superficial landscape but an understanding of the intricate interrelationships of all the parts. Therefore, there is a need for rededication to the undergraduate curriculum and a recommitment of the best faculty to the challenge of undergraduate instruction.

The agricultural system is much larger than it used to be and is increasingly interwoven with the natural landscape and into the full fabric of society. Definition of the domain of agriculture is important for determining what is taught, what is researched, and what the future of agriculture will be. Agriculture is a system of land custodian-farmers and agribusinesses that supply production inputs and process and distribute agricultural products. Consumers, community services, food safety, water quality, climate change, economics, energy, biotechnology, and the environment are all part of the system. Even vocal environmentalists and animal rights activists are part of the system.

However, a number of colleges of agriculture have expanded their curricula even further and have moved beyond the scope of subject matter delineated by the traditional domains of agriculture, food, and natural resources. They have come to serve the aspirations and needs of society as a whole, needs of paramount importance because they deal with the fundamental resources—food, energy, environment, and economies—on which civilization is based.

In defining their priority programs, colleges of agriculture distinguish themselves from other colleges at a university. Education of undergraduates is but one of the roles of the colleges; research and extension join education as justification for agricultural faculties. These colleges should draw on the wealth of their sciences to address the general issues facing society, contribute special expertise to other curricula in the university, and provide a general education to students in colleges other than colleges of agriculture. Understanding of education in agriculture as a new science-based program that focuses on issues of biotechnology, environment, energy, conservation, information technologies, rural communities, new materials, and economics will attract the best students to colleges of agriculture.

These are the challenges for colleges of agriculture, whose graduates must be prepared to address change, constructive conflict, and communication and cooperation among the players: industry, regulators, scientists, extension personnel, farmers, legislators, and the general public. The new model is an integrated system that embodies the basic sciences, their applications, and the markets and consumers of knowledge. In this model, colleges of agriculture are the critical elements in the transformation of knowledge for the benefit of society.

Constraints to Innovation

Changes in curricula and movement in new directions face constraints, some of which are shared by other colleges in the university. A fundamental enigma faces undergraduate education in agriculture, however: People cling to notions of rural pastoralism and simplicity about what is in fact a highly sophisticated system for food production and distribution. This leads to questions concern-

3

ing colleges of agriculture. Will it be appropriate to track undergraduates in agriculture separately from other professional students? Should professional education in agricultural fields be at the master's instead of the bachelor's level? Should there be a separate curriculum in agriculture? Are faculty in colleges of agriculture prepared for the challenge of teaching students with divergent backgrounds?

Another constraint is that graduate students and young faculty are instilled with the view that the path to gaining preeminence is research, not teaching. This is partly the result of the ready accessibility of tools of measurement of and standards for creative and effective research. In contrast, there is an ambiguity of expectations in teaching. Added to these constraints is the fact that faculty feel pressure to obtain extramural grants as a result of the serious erosion of the traditional base of funding for research.

Changes in Curricula

A radical rethinking of the mission, need, and approach to the undergraduate curriculum in agriculture is needed. There is an even greater need to find innovative ways to attract students throughout the university to courses in agriculture, food, and natural resources and for colleges of agriculture to be involved in educating an urbanized society about these areas. The keys to curriculum revitalization are not changes in the courses but changes in course content, goals, and purposes.

Attention to the connections is needed:

• The faculty reward system must be reconnected to teaching.
• Interest in the environment, food, and agriculture must be connected to the teaching of science and technology and social and humanitarian insights.
• Teaching of certain fundamental skills that everyone must master to be able to handle whatever issue with which they are confronted must be connected to the subject being taught.

Industry and Academia

Curriculum development is a process of integration. Just as there is a sense of isolation among colleges and departments, for some years there was also a distance between industry and academia. Substantial changes in industry have transformed agriculture: consolidation of production units, replacement of small communities by regional markets, greater market orientation in government policy, computerization and larger, more efficient production and processing equipment, biotechnology, changes in production systems, the dominance of a limited number of large food system companies, coordinated supply and marketing, and the economic dominance of the transportation-processing-distribution sector. The changes have been so substantial that corollary changes must occur in the academic process, because graduates must be able to manage change, solve problems, make decisions, analyze data, and create new products. However, businesses are finding it increasingly difficult to employ, retain, and reward people to compete in a technology-driven world economy. Recruitment of students and continuing education are needed, and industry has a role and responsibility in both areas. Industry-academic linkages should be fostered and seen on campus. Colleges of agriculture should give attention to the executive potential in students and graduates and should help them to obtain "combat" experience in business, such as through internships.

Needs in Education

The needs in education can be stated in relatively simple terms: Because no one can predict what the issues will be in the next century, no one can determine exactly what courses students should be taking today. Fundamental skills can be mastered, however. These will likely be confidence, motivation, responsibility, effort, initiative, perseverance, caring, teamwork, common sense, and problem-solving and persuasion abilities.

The overriding need for the integrative point of view will require a total transformation of academic thinking and, in the process, the remaking of both higher and lower education. The land-grant university was founded on a sense of place, an integrated landscape with people that needed help. The environment, which is not a subject or discipline or a commodity or resource, can be used as an integrative theme—no discipline need be excluded. Faculty should turn outward from the department, the profession, and the institution. John Gordon suggests that the elements of productivity, sustainable development, environmental ethics, and the application of science and technology can be integrated in a problem-solving course in the first term of college. This can be followed with the requirement that students acquire skills in science, history, and mathematics and knowledge of a foreign language. Then there can be some specialization.

Liberal and General Education

Core curricula, distribution systems, liberal education, and general education are themes that penetrate any discussion of curriculum revision in the 1990s, often as hegemonic aspects. The definition and application of these domains, however, are less certain than the expectations that they ought to be there. The thoughts of

5

Lynne Cheney and Gary Miller provide useful points and counter-points.

Both Cheney and Miller suggest that confusion exists. The terms *core curriculum* and *distribution system* are used interchangeably. So are *liberal education* and *general education*. Miller argues that *core curriculum*, *distribution*, and *interdisciplinary* are terms that reflect the way education is organized, not its content. A distribution system suggests that students should gain broad knowledge and should not be concerned with the specific knowledge they obtain. A core curriculum holds that certain specific areas of knowledge are important, enough so that all students should share them. Interdisciplinary refers to the way in which knowledge is compartmentalized.

Cheney assumes that part of the knowledge gained from the central curriculum should be in the major areas of human thought. This core should ensure opportunities to explore the major fields of human inquiry in broad-ranging, ordered, and coherent ways. She noted that liberal education (traditionally concerned with ideas in the abstract, preservation of universal truths, and intellectual development) assumes that knowledge is important in its own right.

Some institutions, however, do not present a core curriculum but have a rather loosely stated distribution requirement that students take a variety of courses that offer highly fragmented bites of knowledge, without connection. Designing a coherent and rigorous plan of study is difficult, however.

Some institutions have found ways to thwart the need for a core curriculum, arguing that rather than a core body of knowledge, methods of inquiry, approaches to knowledge, ways of knowing, and ways of thinking are more important. Cheney asks what value is learning the ways to know if certain fundamental pieces of knowledge are not known?

Miller argues that a distinction should also be made between general and liberal education. General education is comprehensive; it deals with basic contexts, methods, attitudes, values, and skills that apply in all areas of students' lives. It is not limited to the first 2 years of study at a university but is integrated with the whole curriculum, including professional programs; it centers on the individual student and is concerned with the development of the learner.

Paul Thompson also called for what he termed "targeted social sciences and humanities education" in the curriculum. He distinguished between liberal and general education requirements in the social sciences and humanities (knowledge needed in all university curricula) and targeted education (knowledge and skills needed by agricultural professionals). Discussing the content of targeted education, he noted that both agricultural leaders and citizens need a more sophisticated understanding of the food system, including the social, ethical, and cultural values that relate to the system. Traditional targeted social science education has stressed farm management, agribusiness, and community development. Future education should be provided in a framework that acknowledges that future consumer and political decisions will be made by a society that has little formal or life experience in agriculture. Thompson calls for a nationwide movement that would target social sciences and humanities through faculty positions and initiatives by colleges of agriculture.

New Curriculum Principles

The committee that developed the conference recognized that new principles should guide the construction of curricula in colleges of agriculture. Aspects of these principles developed throughout the conference.

Some participants expressed concern that although recent enrollments in colleges of agriculture have rebounded, many students have shifted to biochemistry and genetics, agricultural business and management, nutrition, and natural resources. Some participants suggested, however, that these are the directions that modern colleges of agriculture should take.

The discussions that focused on cultural diversity were mindful that the major net increases in the work force in the next decade will be predominantly women, members of minority groups, and immigrants. The requirement is for broadened, more inclusive, and relevant curricula. Role models, sensitive awareness of cognitive diversity, mentoring, and financial assistance are key elements in fostering multicultural diversity in colleges of agriculture.

The vision of agriculture has included the vision of good stewardship. This image was fostered in part by the Jeffersonian agrarian ideal, the attention that many farmers of the country have given to the farm landscape, and, perhaps, the cessation of the Dust Bowl. This traditional social and ethical context of agriculture—the special relations with the natural world and the values of labor, the community, and rural life—once regarded as unique can no longer form the basis of its operation or existence. With this assertion, Otto Doering argued that the industrial sector has overtaken agriculture and has permanently altered the context of agriculture for many Americans. Students come now without a sense of context and only a limited set of strong social and ethical norms. On the other hand, scientists have had little interest in such norms. What is needed is a reevaluation of the way agriculture and colleges of agriculture approach the rest of society.

Some participants concluded that faculty members often attempt to separate scientific evidence from values rather than bringing all information to bear on the whole. Debated was the question of whether courses in colleges of agriculture could be used to teach the values and ethics of agriculture. However, some participants concluded that if colleges of agriculture fail to deal with the social and ethical contexts, agriculture would become paralyzed as an industry.

Robert Matthews argued that an environmentally sensitive curriculum should provide both relevant understanding and appropriate analytical skills. To do that, students will need to understand the historical, social, political, and economic contexts within which environmental problems have emerged. The analytical skills needed are problem recognition skills (ability to recognize a problem when they see one) and policy evaluation skills (ability to evaluate proposed environmental policies). These skills presume knowledge of environmental risk and risk management, the policy process, and ethical theory.

Eugene Allen argued that the content of curricula should be based on the needs of the undergraduate as a college graduate who will pursue multiple careers in a lifetime, not as a future professor whose career flows within the confines of a single discipline. Change in curricula can no longer be achieved through employment of new faculty and is more dependent on leadership than on the allocation of resources.

Jo Handelsman suggested that the curricula and ambience of colleges of agriculture are ensnared by visions of the past, even though their science is modern. The field of agriculture takes on the image of being old-fashioned and traditional. These aspects can make courses in colleges of agriculture less attractive to students in other colleges in the university.

It follows that faculty members can contribute individually to a wider use of agriscience and agribusiness courses in other university curricula by presenting the concept that current and future agriculture is rich with the potential for change and improvement, projecting the strengths of the sciences related to agriculture, food, and natural resources. The belief that agriculture can best be marketed as basic sciences was expressed. This may reach for the view that traditional courses in agriculture present only a superficial, mundane look at the whole and choose the basic science route as a means of developing depth and interest rather than transforming the mission of colleges of agriculture.

Precollege Education

The importance of the precollege years in setting an interest in study in colleges of agriculture; the desirability that some understanding of agriculture, food, and natural resources be inculcated in the young; and the special obligations colleges of agriculture may have in facilitating improvement in these factors as well as increasing student interest in science were the focus of several parts of the program.

William P. Hytche sets a high priority on the development of minority human expertise to play a vital role in maintaining a stable professional work force. He is concerned, as many are, of the crises in the education of minorities. He believes that colleges of agriculture can and should lead the way, even in the face of the fact that many minorities still regard the agricultural, food, and natural resource system as the basis for prior enslavement. However, recruitment should begin years before secondary school. Hytche argues for special efforts and nontraditional approaches, first, to encourage minority students to seek higher education and, second, to enhance their interest in and perception of agriculture.

One of the arguments was that colleges of agriculture have special opportunities and responsibilities for intervening in the entire precollege education. The traditional relationships of colleges of agriculture with vocational agriculture and 4-H Club and related youth programs generally remain in place. Colleges of agriculture can and should assist in making the necessary changes to update these programs in the image of the changing views and thrusts that were the subject of the conference. A large problem exists, however, in that students in many school systems have no access to courses in vocational agriculture or agricultural sciences in secondary schools, to knowledge about careers in agriculture, or to projects in primary schools that relate to agriculture and forestry. In fact, many such students have so little knowledge of agriculture that the negative image of agriculture of the past is no longer a factor. Exposure of primary school students to agriculture through such notable movements as Agriculture in the Classroom (a program of the U.S. Department of Agriculture [USDA]); Project Food, Land, and People (a nonprofit, interdisciplinary supplementary education program of USDA emphasizing agriculture and conservation); Project Learning Tree (for students from kindergarten through grade 6 sponsored by the Western Regional Environmental Education Council and the American Forestry Institute); and Project Wild (for students from kindergarten through grade 12 sponsored by the Western Association of Fish and Wildlife Agencies and the Western Regional Environmental Education Council) and linkages to high school teachers of science, particularly biology, are clearly in the self-interests of agriculture and colleges of agriculture.

Scientific Literacy

Scientific literacy, or rather the lack of it, is an issue that colleges of agriculture share with the rest of the scientific community. The general conclusion was that science education at both the precollege and college levels is inadequate, uninteresting, and irrelevant to students' lives. Colleges and universities are graduating students who have little knowledge of their physical world, and working scientists are often scientifically ignorant outside of their own narrow specialties.

The major error that educators make, Robert Hazen argues, is that they try to make the students "miniature scientists." Some work their way through, of course, going on to professional specialized science. The others are left to feel excluded. Hazen suggests a straightforward solution: a course of study for nonscience students that integrates the scientific views of physics, chemistry, biology, and earth science and then explores one subject in greater depth.

Paul Williams also believes that scientists have preempted science from the public by the way in which it is presented. It has been a substitution of knowledge produced by science rather than the process of scientific inquiry. Science courses should find a balance between scientific knowledge and inquiry and should be participatory.

Systems

Answers to the issues of agriculture might be found by the systems approach (a discrete course component or a method of organizing knowledge), in addition to social science or ethical analysis. This approach concentrates on interactions among parts and on properties of systems that are relevant in problem solving. It is a way of doing. Conceptually, it integrates knowledge from reductionist science into a model of the whole.

The quantitative model and simulation, the "hard systems approach," have been widely integrated as a process into agricultural research and education. It was suggested that there be greater use of the ideas and methodology of the "soft systems approach," which can be modeled as a conceptual rather than a mathematical approach.

Innovation will be necessary, as will special training and opportunities for faculty to gain experience. The student body is changing; the intuitive ability to integrate unconnected courses in a curriculum may be decreasing, thus making a systems component important in the new curriculum.

Systems studies can complement studies of social, historical, and ethical contexts. Values studies, however, may have an advantage because there is an identifiable body of research, albeit largely outside of colleges of agriculture, and the support of specific journals in the field. The soft systems approach is new—faculty members are only now beginning to learn of it—and it has little or no published research base.

Positioning Professional Education in Agriculture

How do we educate students to meet the demands of the world? Changes should come out of the faculty; the methods and the materials are already there. Changes will not come easily, however; nor can any past vision of what was great and what was achieved be the standard of tomorrow. The research and teaching amalgam is the pride and heritage of colleges of agriculture, but important adjustments should be made. Redressing the balance between teaching and research should benefit both. Funding and rewards for both is a requirement for progress. There must also be a change in national attitude. All of those people who are involved—faculty, deans, students, chief executive officers, and presidents—must become active to produce educated individuals prepared for sensible planning.

Undergraduate education is in a new era that is post-labor intensive, post-mechanically intensive, and post-chemically intensive. The era ahead, variously described as environmental, biological, information and management intensive, global, and culturally diverse, will be complex. Challenging colleges of agriculture are faculties less knowledgeable of the breadth of agriculture; student bodies with no connection to the food, agricultural, and natural resource sciences; increasingly sophisticated clienteles; serious human resource deficits; and a pervasive international complexity.

Peter Spotts posed the final enigma: People do not always understand science and technology, even though they affect human lives. Although people sense that science is one of the significant trends shaping the future of agriculture, they do not know and are not concerned about the future shape of that science. Academic institutions should cooperate with the public as it tries to make sense of where science will take them and should be sure that their students—even those in specialized disciplines—emerge from their universities well-rounded, able to function in a society that looks on "experts" with a mix of admiration and suspicion, and with a level of scientific literacy that helps them respond intelligently when public policy issues affecting science arise. It is important to train the coming generation of scientists to consider the ethical, economic, environmental, and social impacts of their work. The reason is the continuity of public support for research. That calls for building the case for spending funds, for integrity in the way that funds are spent, and for integrity in the way that science is done. The role of faculty as educators goes beyond the classroom, so that they can teach others the language of science and so that others, too, can understand.

11

Ideas for Change

The discussions in the conference suggested three essential but not mutually exclusive components of undergraduate education in colleges of agriculture:

• the part that builds a core knowledge base: both liberal education and the education providing the knowledge characterizing the agricultural major;

• that part which is problem-based learning (McManus, 1991; Walton and Matthews, 1989), including both the general education that builds contexts, methods, attitudes, values, and skills applicable to all majors and the wide-ranging optional courses and modules that present not merely facts but that integrate and present ideas, logic, and approaches enabling the graduate to manage change; and

• that part which is experiential-based learning and develops the individual: experiential and hands-on education, language skills, interpersonal skills, and specific problem-solving abilities.

Participants offered ideas on how these goals can be accomplished.

Professional education in agriculture should be defined to lead to graduates who have a strong substantive base that underpins agriculture and that is as rigorous and delivered with as much excitement as any other professional course.

As technical content is consolidated and integrated, so should the human elements, such as ethics, literature, philosophy, foreign languages, geography, history, and political science. Courses should provide instruction in concepts, synthesis, and process to make students better problem solvers.

Integrative thinking will be among the requirements of graduates in the future. In the past, colleges of agriculture depended upon the student to take discrete and separate courses and set about constructing their own universe and finding coherence in it. Such students, often with backgrounds in farming and related activities, were able to do so and present themselves as successful, flexible, and adaptable graduates. The abilities for integrative thinking must now come as much or more out of the college ambience.

The construction of a core curriculum should concentrate on both the knowledge content, that is, what should be taught and learned, as well as ways of teaching, learning, and knowing. It follows that problem-based (general) education should concentrate on the problems to be solved as well as the process of problem solving.

Faculty members should openly address the relationships among general education, professional education, and disciplinary specialization. Curriculum and pedagogical approaches should effectively move more disciplines together so that multidisciplinary approaches can be adopted by faculty with a sense of familiarity.

Colleges of agriculture should turn their teaching capacity to the entire food system and to the broader benefits to society, noting that it is in the definition of their priority programs that colleges of agriculture distinguish themselves from other colleges at the university. Colleges of agriculture should open their range of courses, especially in those disciplines that are a part of the agricultural system but that are pertinent to other elements of society, such as business management, personal enterprise, communication, engineering, biological and biomedical sciences, and environmental sciences.

New dimensions are required in the curriculum, many of which are alien to the experience and knowledge base of the faculty. Emergent is an imperative for colleges of agriculture: a greater investment in faculty development. Faculty members need opportunities to learn and develop concepts, techniques, and skills and to think about and plan teaching approaches.

There should be an administrative advocacy and commitment for excellence in teaching, as well as research and service and quality of students. The reward system should be based on a developed philosophy clearly delineated by the university administration. Both teaching and research should be seen as essential and complementary activities. The system should provide incentives for curriculum innovation. Objective evaluations, not only student evaluations, should be ongoing. Complementary peer evaluations should be included. Implementation of a balance between teaching and research should avoid any premise that deficiencies in undergraduate education occur because faculty members are too busy doing research.

Faculty should be evaluated for their scholarly integration as well as research capabilities at the time they are employed. There is a need to improve the ways in which graduate students and young faculty are prepared for their role in teaching, such as the development of well-defined skills in communication, contact with master teachers, short courses in methods of instruction, and supervised guidance.

To improve the efficiency of faculty contributions to teaching, innovative approaches should be explored:

- reducing skill training or how-to courses;
- panel teaching of courses by instructors with different specialties or from different fields to facilitate the integration of multiple inputs of information and merged perspectives;
- ensuring that each adjunct faculty member has a specified teaching role;
- teaching with flexibility with regard to the times and course credits offered;

- encouraging student participation in teaching, with the instructor as guide or mentor;
- revising or expanding the concept of scholarship to include more than scientific research, such as the integration and application of knowledge; and
- emulating the small, successful liberal arts colleges.

The public perception of what teaching is should be broadened. The public should know that a rich diversity of teaching situations exists beyond the classroom. Teaching can be by example, by sharing mistakes and findings of research in the classroom, and by writing.

Colleges of agriculture should be concerned about the issues facing students who are members of minority groups. They should be encouraged to study science. Colleges of agriculture should be active and innovative in recruiting minority students to meet the needs of the future professional work force in agriculture, food, and natural resources. They can lead higher education in this endeavor.

Cognitive diversity is a curriculum dimension of importance. Special attention is needed to accommodate and encourage cultural diversity. Students and faculty also need to be aware of the history, culture, and cognitive styles of members of all ethnic groups.

Colleges of agriculture have the opportunity and a responsibility to extend knowledge about agriculture, food, and natural resources to precollege students at all levels and to their teachers. The mechanisms available to colleges of agriculture may be national projects that provide teaching materials, direct provision of information concerning careers in agriculture to the secondary school system, and direct and indirect linkages with science teachers and their organizations. Colleges of agriculture should be involved in the preparation and distribution of instructional materials to maintain the traditional student pipelines and to improve the substance, content, and format of available materials.

Faculty and administrative support is needed for curriculum change. The key to faculty support is opportunity for research, scholarship, and writing. Courses will develop only as textbooks, anthologies, and research monographs are produced. The process of revising the curriculum should be a group process led by a diversity of respected faculty leaders.

Colleges of agriculture should teach the basic sciences appropriate to agriculture and natural resources, develop courses that explore the interface of society and agriculture, develop teaching that results in the active involvement of diverse perspectives, and attract a more diverse student body to dispel the image of insularity.

Science should be taught by colleges of agriculture and otherwise as inquiry, not simply as a body of knowledge or given answers, and should be a participatory activity involving both teachers and students.

Constructive change should be forged in a partnership among industry, academia, and government. Preparation for careers in business and industry requires a commitment by industry as well as academia.

Colleges of agriculture should attend to the specific social and ethical contexts of agriculture, food, and natural resources. However, colleges of agriculture should include a component of education in the social sciences and humanities, over and above university-wide core requirements, that is specifically targeted to provide appreciation for the historical roots of agriculture and for the social, ethical, cultural, and critical issues related to agriculture, food, and natural resources.

Support mechanisms, including systems research, should be created for experimentation and implementation of systems studies, including the soft systems approach.

Colleges of agriculture should join their universities in their global efforts. These colleges should turn to educating a larger number of students to deal internationally.

References

McManus, I. C. 1991. How will medical education change? Lancet 337:1519–1521.

Walton, H. J., and M. B. Matthews. 1989. Essentials of problem-based learning. Medical Education 23:542–558.

AGRICULTURE AND
THE UNDERGRADUATE

Introduction

FRANK PRESS

A good education is the essential starting point toward the resolution of the current and future issues that we face as a society.

In 1988, the U.S. Department of Agriculture (USDA) and the Board on Agriculture of the National Research Council directed attention to graduate education in agriculture at a conference that honored the first class of USDA food and agricultural sciences national needs graduate fellows. Other studies that have been conducted by the Board on Agriculture at the request of USDA are *Understanding Agriculture* (National Research Council, 1988b), which focused on secondary education in agriculture and agricultural literacy, and *Educating the Next Generation of Agricultural Scientists* (National Research Council, 1988a), which directed attention to the education of and future demand for doctoral and postdoctoral agricultural scientists and engineers.

In 1991, the focus turned to undergraduates. The National Research Council, through the Board on Agriculture, was greatly pleased to cosponsor, along with USDA, the conference Investing in the Future: Professional Education for the Undergraduate, which emphasized the general education of undergraduate students—nonscience majors as well as students who intend to pursue agricultural science careers.

According to several recent news accounts, individuals in all segments of our society are lacking the background of a solid general education. Recently, a reporter asked several graduating Harvard seniors what accounts for summers and winters. They could not explain it. The reporter asked the graduates to explain the difference between a molecule and an atom. They could not do it.

Recently, the chancellor of the University of California at San Francisco told me of a public meeting where local citizens were protesting, on environmental grounds, the opening of a new biology laboratory. The chair of the meeting, a college graduate, said, "We know that you're releasing DNA from this building and we are opposed to it."

All of us must be concerned about scientific and technological illiteracy in our society, and we must be involved in correcting the situation by better preparing the citizenry—policymakers and voters—through undergraduate education and ongoing communication of science to the public. A good general education is a starting point toward the resolution of the issues we face as a society. Decisions relating to science and science policy are being made daily by lawyers; business executives; local, state, and federal policymakers; and others in positions that give shape and direction to standards in society. Therefore, we must consider science education an essential part of the undergraduate curriculum of nonscience majors.

Agriculture is a striking example. Students who are nonscience majors as well as those who are planning agricultural science-related careers must be educated in agriculture, because agriculture extends into every segment of our society. People must be educated to make choices for themselves and their families—choices that might involve an understanding of such matters as recombinant DNA, nutrition, and the assessment of risk.

In turn, professionals working in agriculture must learn to address the scientific and technological problems that are interwoven with issues of social and cultural standards, ethics, and human values.

We are proud of some of the recent accomplishments of the National Research Council in this direction: a guide for the high school curriculum in biology, better ways to teach mathematics to undergraduates in colleges and universities, and the National Science Resource Center, which, in association with the Smithsonian Institution in Washington, D.C., is a national facility for making curriculum materials available, especially in elementary schools.

At the National Research Council, we intend to increase emphasis on the problems of education in the United States. The National Research Council has recently established the Coordinating Council for Education. The 13-member coordinating council will facilitate and coordinate the variety of current studies and activities in education that are being conducted throughout the National Research Council. Continuing the life-long education and retraining of a work force that must constantly adapt to technological change is of increasing concern within the National Research Council, as is the scientific literacy of the general population.

I am delighted that the people who can bring about changes in education—the people who participated in the conference—have been willing to share their ideas in the papers in this volume, and in this way make them available to educators and institutions across the United States.

References

National Research Council. 1988a. Educating the Next Generation of Agricultural Scientists. Washington, D.C.: National Academy Press.

National Research Council. 1988b. Understanding Agriculture: New Directions for Education. Washington, D.C.: National Academy Press.

CHARLES E. HESS

The U.S. Department of Agriculture (USDA) took great pleasure in joining Frank Press and the National Research Council in cosponsoring the first national conference on the evolving mission of the nation's colleges of agriculture, Investing in the Future: Professional Education for the Undergraduate. We are proud that the Board on Agriculture of the National Research Council joined us in this endeavor and took the responsibility for hosting and planning the conference. It is a very appropriate follow-up to the Board on Agriculture's study on research, *Investing in Research: A Proposal to Strengthen the Agricultural, Food, and Environmental System* (National Research Council, 1989), which has provided the foundation for the national research initiative in agriculture, food, and the environment.

Curriculum Issues

In the food, agricultural, and natural resource sciences, curriculum revitalization is essential to the survival of higher education as we know it. Over the years, our curricula have undergone significant changes, from the pre–1970s emphasis on production agriculture to a strong business approach in the 1970s. This was followed by greater attention to the underlying sciences of agriculture in the 1980s, and today, in the 1990s, we find ourselves in a new wave of transformation that emphasizes the educated person and a broader and more philosophical approach to preparing people for life.

In colleges of agriculture and natural resources, we are now giving greater attention to the global perspective: systems models, problem-solving techniques, environmental ethics, social issues, and the critical area of oral and written communications.

Clearly, changes in curriculum content, format, and mission are occurring. Although they are occurring slowly, a new curriculum is emerging. In the process of curriculum reform, however, those of us in agriculture and natural resources must not lose sight of the uniqueness of these technologies and the necessity of maintaining them as vital ingredients of higher education.

21

As we examine the evolution of the curriculum in colleges of agriculture and natural resources, we soon realize that the elements of change envisioned in this volume and at the conference are truly revolutionary. Some of these elements are examined below.

Faculty Development

The faculty in U.S. colleges of agriculture and natural resources hold the key to virtually all that we achieve in higher education. This is particularly true with curriculum change. Faculty understand the complexity of their role. They determine curricula, help mold people for careers, bring a subject to life in the classroom, and generate new knowledge through research. At the same time, technology advances at an unprecedented rate and faculty have less and less time to keep up, much less to get ahead. Faculty development, in addition to administrative development, is sorely needed in our universities. If we are to ensure real progress in curriculum revitalization, time and resources must be provided for in-service faculty professional development. I am proud that the USDA higher education challenge grants and the 1890 capacity-building grants program provide opportunities to fund such programs.

Graduate Students

Although the conference focused on undergraduates, we must also recognize the importance of graduate students in the teaching programs and the major role that many will play as future faculty in colleges and universities.

Graduate students' education in any discipline should not merely be a time to develop research skills and produce a dissertation. They should have the opportunity to teach and to learn the skills of instruction to ensure the perpetuation of effective instruction in colleges and universities. This attention to teaching in graduate student education will do much to help to allay the concerns that teaching and learning are not alive and well at our universities.

Internationalization of Curricula

The conference also addressed the importance of the internationalization of curricula in colleges and universities. Global interdependence, increasing economic competition from other nations, the loss of technological leadership in some areas, and the important role of agriculture in the world marketplace are concerns for all of us. Agricultural business and education leaders

22

have stated their concerns regarding the internationalization of curricula as they view the curricula and research programs in our land-grant universities. We cannot hope to compete effectively until we begin to develop visionary leadership in the graduates of our programs.

Environmental Health Issues

In recent years, agricultural and natural resources technology has been labeled as suspect by many in the media and the general populace. We are viewed by some as having little or no concern for the environment, health, safety, or conservation. It is imperative that we change these negative perceptions by providing students at our colleges and universities with ethical decision-making tools for addressing those issues. We must ensure that our graduates, many of whom will become scientists and leaders, acquire an appropriate sensitivity and perspective.

Multicultural Diversity

Demographers predict that by the year 2000, women, minorities, and immigrants will account for 80 percent of the growth in the U.S. labor force. In its framework for change, the USDA established a goal of building a work force that values cultural diversity. The changing ethnic, racial, and social composition of the U.S. work force, coupled with the need for food, agricultural, and natural resource professionals to work with people from different lands and different cultures, requires that students and faculty become sensitive to the issues of cultural diversity.

Closely linked to this is the importance of markedly increasing recruiting efforts, with scholarship support for ethnic and racial minorities needing financial assistance.

Conclusion

Curriculum development and implementation are complex activities. They are made even more so by the rapid changes occurring around the world. To meet the needs and challenges of present and future shifts, we need well-educated and trained people, and so our educational institutions must change.

This landmark conference offered a unique opportunity for scientists, business leaders, educators, and public officials to contribute to the improved education not only of students of agriculture and natural resources but also students throughout the higher education system.

Reference

National Research Council. 1989. Investing in Research: A Proposal to Strengthen the Agricultural, Food, and Environmental System. Washington, D.C.: National Academy Press.

KARL G. BRANDT

I believe that the principal reason our colleges and universities exist is to educate students. We cannot do without discovery and scholarship, but what differentiates us from research institutes is our responsibility to teach. I worry that the research agenda is driving our priorities, however, and I worry about how we reestablish the essential balance. Our agenda is simple: the education of the next generation of professionals for the food, agricultural, and natural resource system. It is an awesome responsibility.

We live in exciting and challenging times. One of those challenges is the sheer mass of knowledge that we are accumulating, a mass that may occasionally trigger in us a feeling of panic as we confront the necessity of providing our students with a curriculum that is comprehensive and current. What do we include? How do we "shoehorn" into the curriculum new discoveries and technologies? When we add something new, do we take anything out? If so, what? As the technical content of the curriculum increases, must the human content—the ethics, literature, philosophy, foreign languages, geography, political science—decrease toward zero? Logic compels me to insist that the answer is "No." We must remember not to panic, because, as Nobel laureate Peter Medawar observed, "The ballast of factual information, so far from being just about to sink us, is growing daily less. . . . In all sciences we are being progressively relieved of the burden of singular instances, the tyranny of the particular. We need no longer record the fall of every apple" (Medawar, 1984:29).

Do we adopt that philosophy as we plan curricula and courses on our campuses? Our courses should not be exercises in Trivial Pursuit, focusing only on facts and memorization. Our exams should not be multiple choice, because problems in the real world are not presented in that format. We want our students to be problem solvers. Our courses should instruct students in concepts, synthesis, and process. My biases compel me to believe that a college education should be about higher-order learning, about thinking.

We all have biases, and most of the people who participated in the conference arrived with biases and curricular caveats, but without an individual educational agenda. So it is best if we admit at the outset that we have biases. Therefore, before you examine the specific ideas and concepts presented in this volume, I suggest that the reader begin with an immense, invisible canvas, large enough so that the base coat of your pet educational biases forms only a faint background, with ample space to paint in bold strokes, using different colors applied in many layers, as you create a provocative, intellect-grabbing vision.

I hope this picture will not just be a workmanlike copy of what hangs on your campus now, but that it will be vibrant with new ideas. It may harbor some subtle messages (some residue of your biases may still show through, for biases are not necessarily wrong). If we are successful, however, it will be a curriculum worth a thousand words, an educational landscape that will draw students to the gallery of your campus, a work of energy and rigor that will empower your graduates to take on the bright world of the real.

As you contemplate the content of that curriculum, I hope you share a fear of mine. My fear is that a student will graduate from one of our colleges, having majored in X (where X is some narrow discipline), not only believing that she or he will be an X for the rest of her or his life but also, in fact, prepared for nothing but being an X.

Is the goal of an education the engineering of a tool with only one purpose and with only one use? I hope that the answer to that question is "No." Jobs change. People change careers. I believe we must educate our young scholars to be able to grow and adapt, or else we fail in our responsibilities.

I worry about many other things. Do we cringe when we hear our students speak or when we read their papers? Do we force our provincial students out of the comfortable agricultural nest to rub shoulders with students and faculty who do not share the traditional agricultural perspective? Are we educating our students to acquire environmental sensitivity and an ability to listen and respond rationally (rather than emotionally) to viewpoints different from their own? How do we get our students to study or travel abroad, to experience and appreciate diverse cultures? How do all of us come to celebrate cultural diversity on our own campuses to embrace students of color and different ethnicities? I worry about whether we demand enough of our students. All of these issues are discussed in this volume.

But perhaps my greatest worry is this: Do we provide a learning environment that helps our students tie together the isolated subjects that we require them to take, so that they see the connections between the diverse parts of the curriculum and the coherence that we like to believe is there?

This is not a new concern. Nobel laureate Francois Jacob, in his autobiographical work *The Statue Within*, wrote about his early education in France before World War II: "One aspect of the teaching in the lycee bothered me: the compartmentalizing of subjects, the isolation of each discipline. . . . No matter how competent the teacher, . . . the idea never occurred to one of going beyond his

25

boundaries, of showing us that the world is a whole, that life is a composite of many things. . . . It was up to the students to set about constructing their little universe and finding in it some coherence" (Jacob, 1988:59).

Can the teaching of connections become an integral part of our curricula?

Our agenda is simple, but very demanding. We must direct our energies toward composing an undergraduate curriculum that will ensure the molding of minds with voracious appetites for knowledge, minds that willingly see issues from many sides, that have an endless capacity to shape knowledge in new ways, and that are prepared to construct solutions for problems not yet discovered in our changing world.

If I did not care, it would be easy to shrug my shoulders and live with the status quo. But I happen to believe that we can do better—that we must do better—in educating our students.

We must now worry about how to accomplish this simple agenda. I invite you to stretch that canvas I mentioned earlier and open your paint box. It is time to begin.

References

Jacob, F. 1988. The Statue Within. New York: Basic Books, Inc.
Medawar, P. 1984. Pluto's Republic. New York: Oxford University Press.

Part I
Conference Papers

Rethinking Undergraduate Professional Education for the Twenty-First Century:

The University Vantage Point

Nils Hasselmo

In the recently published Carnegie Foundation report, *Scholarship Reconsidered: Priorities of the Professoriate*, Ernest L. Boyer said that he is "beginning to believe that the 1990s may well come to be remembered as the decade of the undergraduate in American higher education" (Boyer, 1990:xi). This thoughtful report should be required reading, because in his usual manner, Boyer offers a rich array of sensible ideas to help fulfill his prophecy.

The conference was extraordinarily consistent with those sensible ideas, and I take that as additional evidence and encouragement that improving undergraduate education has genuine momentum that will, in fact, make a real difference.

I also sensed in the conference a spirit of openness to educational reform without parochialism. Clearly and refreshingly, the conference was not a circling of the academic wagons and a breaking out of the disciplinary rifles. Despite its popularity, that maneuver has never been a good idea in higher education; we usually end up shooting ourselves—or each other—in the foot. It is essential that we look at education in a broader perspective.

The Agenda

What, then, is the agenda in undergraduate professional education from my perspective? What do I see from my vantage point as president of a land-grant university? I hear many voices that say the following:

• Make undergraduate professional education more interdisciplinary.

• Make agricultural sciences degrees more environmental.

• Make health sciences degrees more social.

• Make business degrees more ethical.

• Make engineering degrees more humanistic.

• Make professional degrees more liberal.

• Make liberal degrees more professional.

• Make professional education a master's-level enterprise.

• Make undergraduate education benefit more from the research and public service environments of land-grant universities.

• Make undergraduate education more customer friendly, economical, and effective.

Is this a cacophony of voices? It is certainly a formidable agenda, but it also presents an exciting set of challenges—and opportunities—for constructive change. I would like to comment on four of the many possible topics:

1. The squeeze on undergraduate professional education.
2. The opportunity for undergraduate professional education.
3. The challenge to undergraduate professional education.
4. The model of undergraduate professional education.

The Squeeze on Undergraduate Professional Education

From my vantage point, it seems that undergraduate professional education is being squeezed from below, from above, and from all sides.

From below, undergraduate professional education is being squeezed by the lack of preparation in elementary and secondary schools. How much remedial education can we continue to do and still do our real job?

From above, undergraduate professional education is being squeezed by an enormous expansion of programs at the master's level. As president of the National Advisory Board for the National Study of Master's Degrees, commissioned by the Council of Graduate Schools, I have caught a glimpse of this expansion. There are approximately 800 different master's degrees offered nationwide. At the University of Minnesota alone we offer some 180 different fields and options at the master's level.

From the sides, undergraduate professional education is being squeezed by general undergraduate education in the arts and sciences. As the demands of the professions have grown, the rate of obsolescence of knowledge and the need to teach students how to learn, communicate, and compute and quantify have grown. How liberal must undergraduate professional education be to meet this situation?

The Opportunity for Undergraduate Professional Education

I see considerable opportunities for undergraduate professional education. This is where the land-grant universities, with their broad spectrum of undergraduate (and graduate) professional programs, may have a special responsibility to exercise leadership. I believe those who argue that liberal education can be enriched by the content and the context of the professions and the professional curricula.

I am as firm a believer as anybody in the need for students to understand the basic nature of science and something about the theory and methodology of at least one science. I believe, for example, that students need to study the principles of ethics and the need for social responsibility, that they need some familiarity with our diverse world and our diverse society, that they need historical perspective, and that they need at least some experience with a language other than their mother tongue. Many of the basic questions our students need to understand and try to answer are found in the professions and in properly constructed professional curricula. The skills in communication and computation that are basic to a liberal education are also practiced and taught in specific professional contexts.

In other words, undergraduate professional education can offer content, context, and practice for undergraduate liberal study. Many have said so. We need to keep affirming the principle and steer curriculum development and teaching practice in that direction.

The Challenge to Undergraduate Professional Education

We need to look at undergraduate education more as an integrated whole than as a group of separate disciplines. The future of undergraduate professional education may lie not in an emphasis on professional education but in one on undergraduate education. The future of liberal education may lie not in its avowed lack of professional content and context but in its role in laying a foundation for and creating an understanding of what professional life is all about.

With this approach, undergraduate professional education faces all of the challenges that have been placed before undergraduate education in general and that I listed earlier. With this approach,

31

however, undergraduate professional education can also better serve as a model for solving those problems, which brings me to my fourth topic.

The Model of Undergraduate Professional Education

I believe that we can learn from undergraduate professional education in solving the general problems of undergraduate education. We can learn to insist on *proper preparation* before college work is begun and in working with elementary and secondary schools to ensure such preparation. Professional programs have been better able to establish a distinct set of expectations for precollege students than liberal education programs have. Establishing more distinct expectations is clearly the key to better preparation, and direct collaboration with elementary and secondary schools in establishing the proper expectations is necessary.

We can learn to define the *specific content* of the curriculum and to measure outcomes against established benchmarks. If we cannot define the outcomes of what we do, we are not likely to achieve what we should and will certainly continue to have trouble with our political leaders and the public. The best and only defense against simplistic assessment is thoughtful assessment. It is not easy, but it is necessary. Many professional programs are far ahead of liberal education in definition and measurement.

We can learn to improve *teaching and advising* by structuring the interaction between faculty and students in a variety of ways and by properly using instructional technology. I believe that, in this regard, some undergraduate professional programs have set standards that should be applied more generally.

We can learn to provide *an intellectual context* for the learning and application of theory and methodology that are a central part of liberal as well as professional education. This includes drawing on the richness of the research and public service environment in the land-grant universities, which, I believe, is more common in undergraduate professional education than it is elsewhere.

We can learn to provide *a social context* for learning by integrating the undergraduates into a professional or disciplinary environment. Again, this means drawing the students into a culture of research and public service, that is, a professional culture. Many professional programs do a better job of this than many liberal arts disciplines do.

We can learn to *use, evaluate, and reward faculty* on the basis of all of their contributions to a program and on the basis of specific objectives. Again, some undergraduate professional programs have been pioneers in establishing such systems.

What Are We Doing at the University of Minnesota?

Let me first disavow any suggestion that we have solved all of these problems at the University of Minnesota or that we have done all the things that I mentioned above. We have not, but we have begun to do so, as outlined below:

• We have set new preparation requirements for English, mathematics, science, social studies, and foreign language. The improvement in our students' preparation has been substantial.

• In what we simply call our *Undergraduate Initiative*, we have taken a look at the quality of undergraduate education in our research and land-grant universities. We have put significant resources into undergraduate education through internal reallocation. So far, we have implemented the transfer of almost $5 million into undergraduate education. The plan for the next 3 to 5 years, which was adopted by the Board of Regents in March 1991, involves the reallocation of 10 percent of our tax-funded base budget, with much of the funding going to the units that teach 84 percent of our undergraduates.

• We have seen new, more interdisciplinary curricula, for example, in the Colleges of Agriculture and Natural Resources, that represent distinctive movement in the directions I indicated above. Our general liberal education requirements for students at the Twin Cities campus are in the final stages of revision, with contributions from the professional programs being an important ingredient, with assessment of outcomes being a necessity, with articulation with the secondary schools being a matter of course, and with teaching across the curriculum being an indispensable mode of operation.

• We are gearing up to take a broad look at learning methods and have several pockets of innovation to draw on, several of them in undergraduate professional programs.

• We have developed sessions to determine what makes for a successful department. We are drawing on units that have had great success in national contexts, especially in research and graduate education. Interestingly, but not surprisingly, very successful research and graduate departments also seem to take good care of their undergraduates.

What constitutes a high-quality department that delivers high-quality undergraduate education? I suggest some characteristics below, but with the admonition that the most promising analysis can be had by looking carefully at real departments that seem to have it all.

A high-quality department that delivers high-quality education has a sense of community that is strong and evident. The balance of the mission is understood, and the division of labor is both respected and rewarded. There is a coherent sense of purpose and direction, both for the department and for its individual members. There is a clear sense of quality. There is a clear sense of "customer," even if the term is not the preferred one. In one or more persons there is leadership that knows how to recognize the needs of individuals and that acts accordingly. There is an accountability ethic in which both individuals and the group expect to be judged fairly and rationally. Staff and students are full citizens, and they are treated that way. There is both the security and the strategy to seek out the best and the brightest without socioeconomic, racial or ethnic, or geographical barriers. There is the commitment and ability to develop their talent even more and to integrate them fully into the department and its value system. There is a willingness to target the investment of departmental resources toward clearly defined objectives. There is a sense of departmental community within larger communities. Finally, in terms of the land-grant university, there is special attention to offering the undergraduate the full benefit of the university's research and public-service culture.

These characteristics are anything but the naive wish list of academic reformers. They already exist in departments all over the country, and they reflect a strength in values that will reinforce—not resist—the restructuring and reform that will make a difference. The president who wants to rethink undergraduate education has many talented allies.

Conclusion

In the end, two factors are decisive: culture and economics. The culture of the university represents our values, the values that drive what we do. The culture determines how we implement those values. Faculty members and administrators are an important part of shaping that culture. The economics of the university determine in fundamental ways how much we can do and how fast we can do it. We face economic problems, no doubt, at the national and state levels. If the culture is one of constructive change and accountability, however, I believe that our culture can overcome our economics rather than vice versa.

Reference

Boyer, E. L. 1990. Scholarship Reconsidered: Priorities of the Professoriate. Princeton, N.J.: The Carnegie Foundation for the Advancement of Teaching.

Rethinking Undergraduate Professional Education for the Twenty-First Century:

The Public Policy Vantage Point

Ray Thornton

I have benefited tremendously by my association with people like those who participated in the conference, with people on the campuses where I have served, and with people who are concerned and interested in making our nation move forward strongly. Recently, one of my colleagues in the U.S. Congress asked me, "Why in the world would you leave, being president of the University of Arkansas, to come to Congress?" I said, "I got tired of all the politics." Faculty members and administrators alike recognize that campuses are not immune from politics. As Henry Kissinger said, "The reason that university politics is so fierce is because the stakes are so small."

One of the things that we must constantly keep in mind in our universities, in agriculture, and in our nation is that when we divide the pie between competing groups, we draw artificial lines of demarcation. As the president of a university, I never could understand exactly where the line between chemistry and biology existed or why the battles were so fierce between those on either side of the line. Occasionally, we should step back from the arena of our daily contests and imagine what we might be able to accomplish if we would set out with a new beginning to construct a different fabric to meet today's challenges.

In 1975, Congressman Jim Symington (D.-Mo.) and I chaired our committees (Subcommittee on Science, Research, and Technology and Subcommittee on Domestic and International Scientific Planning and Analysis, respectively) in a series of hearings on the nature of U.S. agricultural research, what direction agricultural science and technology should take to meet emerging needs, how well our system was responding to national needs, and what new means could be used to improve the agricultural research system.

Orville Bentley, former U.S. Department of Agriculture assistant secretary for science and education, was one of those who testified at our early hearings. He said, "The system works, but it is incumbent upon us to make it work better" (Bentley, 1975:52). Out of those hearings came the idea of the National Agriculture Research, Extension, and Teaching Policy Act of 1977, which I introduced in the U.S. House of Representatives and of which I was the principal Democratic sponsor. The 1977 legislation contained measures aimed at stimulating creativity and innovation in research and the application of knowledge. For the first time it set up competitive grants in agricultural research, and it opened the door for agricultural research to be performed by nontraditional sources.

Fourteen years later, we are faced with great opportunities and challenges to forge constructive change in a partnership between industry, academia, and government. These new and complex challenges have developed, in part, because of the remarkable changes in the world in the past few years. With the end of the Cold War and the truly remarkable military victory in the Persian Gulf, it is time that we develop broad, new, comprehensive strategies to accomplish our national goals.

From every success there emerges the challenge of a greater struggle. After World War II, in a remarkable display of altruism and self-interest, we realized that for the security of our nation, Europe, including our former adversaries, and Asia, including Japan, must be restored to economic vitality and health as a bulwark against the fear that the Soviet communist system was going to sweep across Europe and, indeed, the world.

With a massive economic investment approaching 2 percent of our gross national product, we developed the Marshall Plan for Europe and the Truman Doctrine for Asia. The Marshall Plan was not just an application of money to solve a problem, however; it was a stimulus to those countries to do what they could to rebuild their infrastructures, to educate and train their young people and work force, to develop modern technologies and manufacturing skills, and to reinvest their own resources in themselves. It worked well.

The Marshall Plan did not just provide economic help but was buttressed by the continuing dedication of U.S. power to ensure that the communist system did not break out of our containment policies to sweep across the world. We knew that wars and victories on the battlefield might become a substitute for economic self-sufficiency within an aggressor nation. So we contained the threat of Soviet conquest, and the two competing economic systems met and locked in competition. Finally, the Western system of free enterprise and individualism overwhelmed the state-managed economic system. Free societies demonstrated that they could succeed in the complex world in which we live.

We are at that kind of historic moment today. For some months now I have been calling for a Marshall Plan for America, because it is time that our leaders carefully analyze our needs and, rather than continuing to build upon the prior traditions and structures of the past, develop a new way of thinking about the world and addressing its current challenges.

Agriculture is a very important part of that plan, and undergraduate education is a vital part of that new set of strategies. These strategies are not a product but a process of thinking in which we carefully analyze what is required to accomplish three important goals: maintaining our military and economic strength as a foundation for the lasting values of human dignity, responsibility, and democratic ideals.

First, we must keep America the most mighty nation in the world militarily, recognizing, however, that military might can be attained with high technology and airlift and sealift capabilities and does not require us to place hundreds of thousands of foot soldiers across Europe in order to defend it. In analyzing the needs of the United States, the defense of Western Europe is bound to be addressed. In fiscal year 1990, the United States spent $135 billion defending Western Europe, but against whom? The Warsaw Pact has faded. I do not suggest that we withdraw from the world. We must participate in the world, but we must also attain the strongest economic base that this country can have as we enter the next century. On a strong economic and military base we can achieve our goal of being the greatest nation with regard to the human values that are central to our democratic ideals.

Joseph S. Nye, Jr., Admiral Bobby Inman, and others have analyzed America's resources and strengths, and they report that there is no doubt that, properly employed, the United States can be the dominant economic force in the world and can benefit from partnerships not only within this country but throughout the world. The appetites of less developed countries for agricultural goods will increase as they are stimulated to develop their economies. It will be very beneficial to U.S. agriculture for us to recognize the need to stimulate less developed countries to progress toward their own economic development.

It will also be very important to recognize that energy is a vital part of our nation's future. Eighteen years ago, in April 1973, I addressed the U.S. House of Representatives with my first major speech in which I outlined the hazard that we were facing from not having a policy for alternative fuels and from not developing the energy resources we needed to attain a certain level of energy efficiency. I mentioned such things as fuel from grains and shale oil and solar and geothermal energy.

Mike McCormack (D.-Wash.) and I introduced the first six energy-related pieces of legislation to come into the U.S. House of Representatives, but they were ignored (this was several months before the Arab oil embargo of 1973). Everybody was saying, "Hey, we don't have to worry. We're in good shape in this country. All we have to do is buy more." And then October of 1973 brought a stark recognition of our vulnerability.

Many people remember the lines that developed in front of filling stations all across the United States as the oil embargo put a crimp into the domestic economy. Everyone said, "Why haven't we thought of this? Why haven't we done something?" They found, indeed, that Mike McCormack and I had been talking about the energy problem. That is one reason why we both became subcommittee chairmen at the beginning of the next term of Congress, because we had isolated that problem and had begun to address it. Then, the nation began to address it. We developed strong programs for energy independence. And then the oil-exporting countries got smart: They lowered the price of oil. Because of the free market, we started buying more than we ever did before, and as a result, we are much more heavily energy dependent today.

It is time that we consider what measures we should employ to make this country energy independent. Agriculture can play a big part, not only in the use of grain for alcohol production but also in providing the green solar energy converters—the crops—that are essential in making value-added agricultural products. New value-added products will be critical in replacing those that require extraordinary amounts of energy to produce.

It is also vital that the agricultural system maintains a strong role in the United States. For years our policy has called on the nation's agricultural system to provide an abundant supply of clean, pure, and good food, and we have benefited enormously by the work that has been done to provide that food. Today, the rest of the world is catching on, just as they did during the Marshall Plan after World War II, when they learned how to outbuild us in many areas of high technology.

I am deeply concerned about the shift in competitiveness that is occurring today. In many areas we will have the dominant position, but in looking at the trends, as our congressional committees are doing, one can see that the U.S. position is slipping. In materials research, we are out of the game in the area of computer chips. An American invented high-definition television, but no U.S. company can build television sets based on this technology without some help and partnerships with companies in other countries.

We not only need to move our people into better jobs through manufacturing opportunities but we also need to discover new ways to harness the skills and abilities that our agricultural scientists can use to help the United States use its productive capacity to overcome the unfavorable balance of trade that we have had in recent years.

You will understand my parochial reason—coming from Arkansas, which is the largest producer of rice in the United States—why I feel upset when we cannot even put a box of Riceland rice (Riceland Foods Inc., Little Rock, Ark.) on exhibit at a Japanese world's fair. Yet, I do not know how many products that people in Arkansas and throughout the nation buy from Japan. There needs to be a more level playing field on which we are allowed to compete.

Finally, it is important to think about what the role of government should be. Clearly, the role of universities is vitally important, and I think we should listen well to the words that Justin Smith Morrell, a member of the U.S. Congress, wrote in 1859, prior to the 1862 adoption of his legislation. He said, "The modern achievements of skill, enterprise, and science, new ideas with germs of power must be recognized and diligently studied, as they have brought and will continue to bring daily competition which must be met. If the world moves at 10 knots an hour, those whose speed is but 6 will be left in the lurch" (National Association of State Universities and Land-Grant Colleges, 1987:1).

Undergraduate education in agriculture is absolutely essential to any strategy of meeting the new forces of world competition. We should recognize that government's role is limited, but it is important. Timothy S. Healy, former president of Georgetown University, said, "Great universities are not made by governments, they are made by learned men and women who are free to think and dream and by bright students who are free to learn. Government's first obligation is to trust that freedom, and its second is to help nourish it." I subscribe to those thoughts, although there is a reciprocal duty upon a university to be accountable and trustworthy in the application of public funds.

Government has a role similar to that of agriculture and industry: to trust our institutions and individuals and encourage them to become more competitive. I sometimes think that in this country we are like Gulliver, bound down by thousands of threads on the beach at Lilliput. Our free enterprise system and our agricultural capability are the greatest in the world, and yet, we sometimes find ourselves harnessed and shackled by long delays. It is more speedy today for a U.S. inventor to license his or her invention overseas and watch it come back onto the market in this country than it is to go through the hurdles that are placed in the way of an inventor in the United States.

We need to find ways of letting government help to clear the line of scrimmage, to "block" for U.S. agriculture and U.S. business, to spot some particular demonstration projects, and to announce to the world, "We are coming through this hole." It might be high-definition television. It might be fiber optics. It might be the development of a new strain or a new germplasm that will provide U.S. farmers with the opportunity to be more competitive with farmers in other countries. Then we establish a base to succeed in those efforts.

In the United States, we have a great opportunity to harness our inventive genius to the marketplace. That is a role that universities are well suited to doing, through the patent legislation that gives inventors some access to the fruits of their own inventions. We should find ways of persuading U.S. companies that it is all right to work with each other in limited areas in order to put U.S. workers on the line building products for sale around the world. A lack of such leadership encourages, for example, our automobile manufacturers to enter into such partnerships with integrated manufacturers in Japan or Germany. We need to do some careful thinking. There is no simple solution. H. L. Mencken once said, "There is always an easy solution to every human problem—neat plausible, and wrong" (Mencken, 1949:443).

The problem that I am describing is not a simple problem, but our resources are equal to the task. Those of us in government should step back from the traditions and structures of the past and grasp the marvelous opportunity that is ours in the closing decade of this century. We need a Marshall Plan for America—a comprehensive and interconnected strategy to employ all of our best skills and abilities to make this country militarily mighty and economically strong, so that we may continue as the government that provides the greatest degree of individual freedom, competitive education for all, and the other elements that make it possible to protect and respect the qualities of human dignity.

References

Bentley, O. 1975. Agricultural Research and Development: Special Oversight Hearings. Committee on Science and Technology. Washington, D.C.: U.S. Government Printing Office.

Mencken, H. L. 1949. A Mencken Chrestomathy. New York: Alfred A. Knopf.

National Association of State Universities and Land-Grant Colleges. 1987. Serving the World: The People and Ideas of America's State and Land-Grant Universities. Washington, D.C.: National Association of State Universities and Land-Grant Colleges.

4

The Challenges for Professional Education in Agriculture:

A Corporate Vantage Point

Robert M. Goodman

As a former corporate executive, I returned to academia because of a desire to contribute to the role that I believe our colleges and universities must play in the future of agriculture. I have interwoven four themes into this discussion:

1. There is great leverage in seeking solutions to the many problems that we face as a society and in how we handle the undergraduate curriculum.

2. A radical rethinking of the need for and approach to the undergraduate curriculum in agriculture is needed to meet society's future needs.

3. An even greater need lies in educating students in other curricula about agriculture.

4. We will be successful in opening up agriculture to our society only if we broaden the interest and appeal of agriculture to people of all backgrounds.

A major feature of the U.S. agricultural system today, and likely in the years to come, is that it has increasingly become a partnership among public, political, private, and corporate interests. Agriculture's interests, like society's interests in agriculture, are no longer primarily in the public and political domains, as they were when the land-grant movement started.

Few if any other sectors of the U.S. economy are like agriculture, which embodies many of the future imperatives we face as a nation. We have little choice but to design and execute a strategy to maintain and enhance our capacity as a world leader in agriculture.

There are many reasons that favor our success in designing and executing this strategy. Although the United States is a highly diverse and now a predominantly urban and suburban society, it occupies a highly productive and versatile piece of the globe's agricultural real estate. By international standards, for example, in comparison with Europe or Asia, our rural areas are under less pressure from population or other competing uses for the most productive acreage.

The United States continues to lead the world in the development of knowledge and the diffusion of the technologies and know-how that are used in agriculture. The creation of a complete system of research, instruction, and public service that was the great social invention of the late nineteenth and early twentieth centuries and that advanced the perfusion of science into agriculture is largely responsible for this leadership. The land-grant ideal has since become the model for international development that, for example, laid the foundation for the success of the so-called green revolution in countries such as India and the Philippines. It also provided a context for the growth of a strong private sector in agriculture.

Our history as a nation and our shared values have their origins in and continue to derive their strength from agriculture. For all of its troubles of recent decades, U.S. agriculture is still a dominant sector and an important countercyclical factor in the U.S. economy. Thus, we look to the future from a position of strength, of technical and resource superiority, and with history and a certain sense of destiny in our favor.

Agriculture faces enormous change. The problems faced and the solutions provided by our public institutions of agriculture in the past are in many ways very different from those of today and tomorrow. Our society and our world are also different. When we turn to education in and about agriculture, we find particular cause for grave concern about the demands we face simply to keep up with the needs of society in an uncertain future. So we must ask what changes we should expect of our institutions and our people to ensure that we will be as successful in the future as we have been in the past.

To understand these needs and changes, we must begin by understanding the present and future role that agriculture may or will play in our future as a nation. This is not a simple analysis, because U.S. agriculture is itself at a crossroads. If it is to survive and contribute in a major way to our future economic vitality and to fulfill its potential to address in the long term the food and energy needs of the world's population, it, too, must change.

U.S. agriculture will be driven by two major trends that extend well into the past and that reach well into the future. These trends arise from an increasing command over the genetics of our crops and livestock and an increasing sophistication of our approach to agriculture as a managerial activity. Agriculture will be much more biologically and managerially intensive. Nutrients and crop production chemicals will be used prescriptively rather than prophylactically. Among the greatest opportunities for agricultural research and development in the next generation will be work that focuses on agricultural inputs derived from genetic and managerial improvements. The paradigm shift in agricultural production practices is toward biology and away from synthetic chemicals. The demands that this shift will place on professional education for agriculture, even in the strictly production sense, are great, and I suspect they are only vaguely appreciated by even the clearest thinkers.

We are likely to turn more to agriculture for the industrial starting materials that we now obtain largely from petroleum. Eventually, agriculture will also be the likely source of large amounts of our fuel.

We are likely to see more of our agricultural exports be value-added products made at home from commodities grown here and to see less export of the raw commodities themselves. Increasing concerns about the safety of foods and better understanding of the link between food characteristics and nutrition will likely result in stronger linkages between production practices, crop and livestock genotypes, food processing, and marketing. An already technology-rich food supply system will become more sophisticated. The same will likely be so for fiber and forestry.

At the same time that we are dealing with these changes, we must also deal with the need to preserve and enhance the resource base and protect the environment. For a variety of reasons, these considerations probably mean that the amount of land committed to agriculture in the future will not significantly increase over that in use today. Environmental considerations drive significant major needs and opportunities for technologies in the future.

If I am right about the future, U.S. agriculture must become more global. Today, we have companies that operate from our agricultural base in the global marketplace. But the export of, for example, consumer products based on agricultural ingredients is a very different challenge, requiring a level of cultural, social, and political knowledge that is atypical in U.S. agribusiness today.

We must also think about our own markets. Our recent history in the global automobile marketplace, or closer to agriculture, in tractor manufacturing, shows that we failed to understand foreign markets. We also failed to understand our own markets, as the prevalence of imported vehicles and machinery testifies.

What do these speculations about the future of agriculture say about the changing needs in education? To talk of a need for change is not to disparage the past. We have a distinguished history in agriculture and agricultural education. We are not called upon to defend or apologize for the past. The point is not that the past was not good enough but that it is not necessarily a model for the future.

Many have already realized the challenge (Bodner, 1990; Heller, 1987; Koshland, 1991; Nature, 1991), and so this discussion does not describe the situation in all institutions or for all courses, curricula, and professors; some welcome and important progress is being made in some places by some dedicated and sage individuals. Progress in some quarters is merely a reminder, however, of how far the rest of us have to go. It also provides some assurance that the effort will be worth it.

We face significant odds and must work against a background that is broadly and sadly discouraging. Across the board, enrollments in colleges of agriculture are declining. Moreover, we have seen a serious decline in the appreciation for and understanding of agriculture by the educated populace at large. This decline extends, ironically, to many professionals in industries and other activities that are dependent on (e.g., medical care or transportation), if not directly part of (e.g., banking or food processing), the agricultural sector of the economy.

In the United States, people cling to notions of rural pastoralism and simplicity about what is in fact a highly sophisticated system for food production and distribution. These notions are perpetrated and embellished by advertising and political campaigns and, more insidiously, by a naive (at best) and at times seemingly conspiratorial (at worst) silence in our schools. As a result, many people abstractly accept agriculture as a necessity, but their concerns focus on agriculture as an environmental and economic enemy of the people.

Between 1982 and 1990, I served on the leadership team of a successful agricultural start-up company. During that time, I hired over 300 individuals to work in a range of positions from senior scientist to patent attorney and in areas from greenhouse operations to agronomy. During those years, I learned that higher education in agriculture has been marginalized to the point of being nearly superfluous in many areas of modern agriculture. This marginalization is because the education and training offered in many of our colleges of agriculture have not kept up with the forefronts and, therefore, with the basic skills needed in today's job market.

A decade ago I noted that computers seemed to be more prevalent on farms in Illinois than they were in the classrooms of the college of agriculture at the University of Illinois at Champaign-Urbana. Today, those of us who hire research technicians in the private sector find young people with the skills and experience we need as often as not among graduates of programs in chemistry, life sciences, and chemical engineering, and we must absorb the cost of teaching them about agriculture on the job. This is not bad, because in my experience such people are quickly attracted to the importance and the intrinsic interest of agricultural research and development. Both of these observations raise the question of the relevance of having an undergraduate curriculum in agriculture.

Given this perspective on the present predicament, how do we go about putting agriculture back on the main agenda of the people of the United States? The issue comes down to how we educate our college-age population. Our highest leverage as a society is on the 4 years of undergraduate education. The highest proportion of future community, business, political, and intellectual leaders, as well as teachers of students from kindergarten through grade 12, share the experience of these 4 years of learning and growing. It comes at a time of human development in our culture in which most individuals are most open to new ideas and are most intent on an individual search for meaning in their lives. At the same time, they are acquiring skills and experiences that will stay with them for the rest of their lives.

The challenge is to define how the professional educational experience in agriculture in the coming decades will lead to graduates who have a strong substantive knowledge base that underpins agriculture—biology, chemistry, mathematics, sociology, economics—that is every bit as rigorous and delivered with every bit of the excitement of professional courses in other areas, such as engineering, business, law, and medicine.

The discussion must start with consideration of the need for a separate curriculum in agriculture. One must ask at what stage in the preparation of undergraduates is it appropriate to track undergraduates in agricultural subjects separately from other professional students? Consider seriously the question of whether we now do it too early and whether we make the distinction too distinct. Should professional training in agricultural specialties be at the master's level instead of at the bachelor's level?

We should not stop at the organizational questions. We must also ask the hard questions about teaching (Bodner, 1990; Rigden and Tobias, 1991). For example, what about course content? We share with the rest of science and engineering the criticism that our courses are dull and therefore difficult, that our disciplines are authoritarian, that we have succumbed to tyrannies of technique, and that we mistake weaknesses in our pedagogy for a lack of interest or even talent in our students (Tobias, 1990).

At the same time, our educational system must incorporate information and awareness about agriculture into other professional curricula and into the 4-year undergraduate curriculum in general. The challenge is to think creatively about how to make this happen. This is not a call for better public relations. It may be a call for more joint course offerings, but maybe even more radical thinking is needed.

What about having professors in colleges of agriculture teach courses in education, law, medical, and business schools? What about the links between colleges of agriculture and colleges of liberal arts? And do not forget the smaller, nonprofessional liberal arts colleges. Some of the more successful professional curricula at the postbaccalaureate levels have established strong ties to these liberal arts colleges. Can agriculture afford not to do the same? We know from the past that these kinds of links can serve agriculture very well over many years.

A further challenge is to consider how the graduates of professional curricula in agriculture will acquire or maintain and strengthen their personal and professional commitments to the traits that make for a life of learning and enrich the human dimensions of their professional lives. As I think about the needs of industry in a changing world, I think of needs such as flexibility, diversity, perspective, and values.

These needs cannot be fully met by paying attention only to what we teach. They must also be embedded in our institutions and in the way in which we teach (Tobias, 1990). They must be part of the fabric of our student's lives. They must be lived to be truly learned.

Pedagogy is almost certainly part of the answer. Students will learn flexibility if they are taught problem solving. Students will learn the value of teamwork if they are taught in a cooperative learning environment. They will find excitement in their studies and maintain an interest in the difficult courses if they are taught in a way that allows and requires them to discover rather than merely to memorize.

Students will acquire a healthy perspective—on work, on their role in society, on their society's place in the global picture—and at the same time shape their values for a lifetime of true citizenship if they are not just taught about but get to experience in a meaningful way the world that they will become part of when they take jobs. Students in any professional curriculum need experiences in the real world. This experience can be gained in many creative ways, but it will be a step forward just to have more accessible opportunities for the tried and true approaches, for example, through internships, semesters abroad, and exposure to examples set by professors whose experiences include work in industry or abroad as well as academia.

My experience running intern programs with undergraduates, and even some high school seniors, in both industry and the university is that carefully chosen students can fit into and learn an immense amount from surprisingly challenging roles in the real world. The National Science Foundation's Undergraduate Research Participation program of the 1960s is another example of a very simple but effective model for students headed for advanced degrees. I know many professors in agricultural fields today who, like me, had their first taste of research, and for many, their first exposure to agriculture, because of this program.

In industry, we need graduates who have begun to acquire a maturity about the world and their place in it while they are still in the learning environment. We must create or adapt learning approaches that incorporate discovery, cooperation, exposure to the real world, synthesis, and problem solving. One of my favorite ideas is the use of decision-case analysis. The need for maturity again raises the question of whether the appropriate focus for "professional preparation" is at the bachelor's or the master's level.

Diversity is a particularly difficult issue for those of us in agriculture. Quite apart from the legal and moral obligations that companies must meet in their hiring practices, our experience shows that a company with a rich diversity—in cultural, ethnic, socioeconomic, and gender terms—is a more appealing place to work and yields a more productive, creative work force.

No one can structure a diverse work force by recruiting strictly from the ranks of graduates from colleges of agriculture today, because the makeup of that pool of potential employees is impossibly monochromatic. Some progress in the level of gender diversity in some subdisciplines of agriculture has been made in recent years, but there is far to go. In other measures of diversity, agriculture lags far behind, even in what it is trying to do, compared, for example, with minority recruiting and retention efforts at the undergraduate and master's levels in colleges of engineering.

My experience as an employer suggests to me that there is nothing intrinsically wrong or unappealing about agriculture to minorities or women. There are still many barriers, however. These barriers have a lot to do with the kind of environment those of us in the field of agriculture create, often unwittingly, toward those who are interested but who must be shown some enthusiasm and some willingness to accommodate new perspectives and different values and who may require new or different teaching approaches for the most effective learning. We can learn a lot about the requirements of recruiting and retention of minorities and women in agriculture by paying attention to the spirited attention this issue is getting in other previously men-only disciplines as well as from the creative work of cognitive and social scientists (Kolodny, 1991).

The issue about diversity in agriculture is therefore to define how we can specifically recruit and retain more minority and women students in colleges of agriculture. Encouragement, directly and indirectly, of students will be part of the program, but it will not be sufficient. Faculty awareness programs that deal with issues that range from the obvious to the subtle and projection of these awareness programs into the agricultural production and agribusiness communities will be necessary. If we are successful, the jobs will be there for those students lucky enough to be attracted by the programs that we should be developing.

I want to discuss briefly the issue of teaching teachers. It is in the undergraduate curriculum where we have the greatest leverage to address the pressing issues that we face as a society regarding

47

the quality of our precollege (kindergarten through grade 12) system. My conviction is that preparing teachers to teach is too important to be left to colleges of education. Those of us in agriculture need to work at two levels. We need to get behind the programs that are providing materials for teachers already in service to learn about agriculture. Two notable programs are the U.S. Department of Agriculture's Agriculture in the Classroom program and Project Food, Land, and People, a nonprofit, interdisciplinary supplementary education program emphasizing agriculture and conservation. However, it is just as important to engage the state agencies that govern syllabuses and our colleagues who design curricula for training teachers to infuse agricultural topics and hands-on agricultural experience with plants, domesticated animals, forests, food, and the environment into the learning experiences of future primary and secondary school teachers.

The suggestions I have made are not likely to happen without a rededication of our major colleges of agriculture to the undergraduate curriculum and without a recommitment of our best faculty to the challenge of undergraduate instruction. I question whether all or even any of this teaching should be done in the traditional framework of an "agricultural" curriculum. In fact, I have very serious doubts about the typically separate agricultural curriculum, at least for the first 2 years of undergraduate work. I might rather see our professors in colleges of agriculture "professing" in the core curricula of our colleges of arts and letters and, in return, have our upper-level curricula in agriculture draw heavily on some "professing" by professors of law, business, behavior, humanities, and the sciences.

Finally, I want to issue a challenge to my friends and colleagues in the corporate world. They stand to be the beneficiaries of whatever good comes out of the ideas presented in this volume and the rethinking about our educational enterprise in general that these ideas represent in agriculture. I believe that individuals in the corporate world also have a responsibility to make it happen.

The problems in our precollege educational system to which some local communities and their corporate citizens as well as individuals are now awakening require action on at least two levels.

There is much that corporations can and should do at the local level, for example, undertaking local action to support schools and the status of teachers in the community. They can throw generous support behind the Agriculture in the Classroom program and Project Food, Land, and People and create other such initiatives. They can lobby and support the creation of agriculturally related programs in local and regional science centers and be generous supporters of those centers in their outreach programs.

Equally important and, arguably, more highly leveraged opportunities will be at the undergraduate level, however. There is much that companies can do at the undergraduate level, too. They can develop internships for students and mentorship programs for their teachers. They can offer continuing education opportunities for teachers that are linked to the last year of teacher training programs in local or regional colleges. They can offer sabbaticals for their employees; specifically, this will allow company staff to work in places where education of undergraduates in agriculture is a priority, thereby giving professors and students exposure to industry and people in industry exposure to the academic environment where future employees are being educated. They can support generously the recognition of great teaching. They can incorporate explicitly into their gifts and grants for research support funding for undergraduate participation. And they can lend their considerable political support to the challenge of redesigning the curriculum. Wise educational institutions will welcome the support, both financial and political, and the wisest ones will also engage companies in the process in some meaningful way.

The students most needed in agriculture today are those who are most likely to be able to think globally, to act creatively, to value diversity, to behave responsibly, to respond flexibly, and to interact cooperatively. Open and fertile, inquisitive, and observant minds, not "finished products," should be the goal.

In society at large, we desperately need a higher level of general science literacy and specific understanding of agriculture. In today's world, however, the need is as great to equip all people in our society with an understanding of agriculture. The need is not limited to just more facts and figures about and exposure to agriculture. A better appreciation of probability, for example, would equip people with the ability to better understand risk. Thus, one might argue that a better understanding of applied mathematics could do more to advance people's understanding of the implications in their lives of agricultural production practices than a better understanding of agricultural production practices themselves could.

The future of U.S. agriculture in no small measure rests on how well we meet the challenge of placing education in and about agriculture back on the main agenda of our society. If the conference and this volume collectively make some progress in moving in this direction, we will have done something worthwhile.

References

Bodner, G. M. 1990. Falling grades for college-level science. Chemical and Engineering News 68:69–70.

Heller, S. 1987. Ways to improve undergraduate education sought by alliance of state universities. Chronicle of Higher Education, January 14, 1987, pp. 13–14.

Kolodny, A. 1991. Colleges must recognize students' cognitive styles and cultural backgrounds. Chronicle of Higher Education, February 6, 1991, p. A44.

Koshland, D. E., Jr. 1991. Teaching and research. Science 251:249.

Nature. 1991. Education in science. Nature 350:3. (Editorial.)

Rigden, J. S., and S. Tobias. 1991. Too often, college-level science is dull as well as difficult. Chronicle of Higher Education, March 27, 1991, p. A52.

Tobias, S. 1990. They're Not Dumb, They're Different: Stalking the Second Tier. Tucson, Ariz.: Research Corporation.

5

The Environmental Curriculum:

An Undergraduate Land-Grant Future?

John C. Gordon

These are heady times for those centrally concerned with the environment, but they are hard times for universities. In the former category, we have the atmosphere and forests emerging as global and foreign policy issues at par with war and peace. On the latter front, we have books entitled *Profscam* (Sykes, 1988), *Killing the Spirit: Higher Education in America* (Smith, 1990), and *The Moral Collapse of the University* (Wilshire, 1989). Linking the two themes, David W. Orr, writing in *Conservation Biology* (Orr, 1990), asks, "is conservation education an oxymoron?" Land-grant universities are getting their share of opprobrium for not being environmental or sustainable enough and for being obsolete in an increasingly urban nation. I argue that no amount of curriculum tinkering and academic committeeing will cause us to rise from these doldrums and to the environmental occasion. We must, rather, transform ourselves entirely and, in the process, remake higher and lower education, particularly the part of it concerned with what we call science education. This is entirely in the land-grant tradition. We invented ourselves once; we can do it again.

What to Build On

Land-grant universities were founded on a sense of place: an integrated landscape containing people who needed help. They were directed at altering the environment for the better, with better farmers and mechanics as tools. They realized that people were at once the primary problem and the principal resource. They set out humbly and realistically, but with a massive sense of purpose and destiny. If we can rediscover those attributes, we have all we need to build on.

The Problem

Suppose that, for once, the popular perception is correct: It is time both to make environmental concerns central in human affairs and to seriously change universities and schools in the United States. If that is so, there is a striking symmetry. The major barrier to making the environment functionally central is that we know so little about it. Not only is research deficient (see, for example, recent National Research Council [1989, 1990] reports) but people have not been taught much about environmental questions and solutions. On the other hand, universities have avoided making the environment an important component of their "product mix," except when it is a trendy topic, and even then their main contributions have tended to be rhetoric that warns of environmental problems and that attempts to establish primacy in the environment for a particular faction or unit of the university.

In their current configuration, universities do not usually deal well with the environment, because they must serve disciplinary probity. Because the environment is everything, it tends to have no constituency and, thus, to be functionally nothing to universities that demand disciplinary packaging. Some parts of some universities package things by professions, commodities, or "resources," which has essentially the same effect. The environment is not a subject or a discipline, a commodity or a resource, or even several of these. It is, rather, an integrating point of view. Indeed, it is the most integrating point of view possible. In perhaps its simplest form, it is the point of view of those who see that ensuring the livability, productivity, and beauty of the earth is not only an important task but is the central mystery and, thus, is worthy of the most rigorous scholarly attention.

It is also the single most obvious organizing principle for places with the temerity to call themselves universities. Therefore, the first problem for the "environmental university" to solve is to keep the environmental point of view and the studies that this calls for from the existing conflict and competition with the various disciplinary departments and professional schools. This is probably best done by adopting an open, external problem-solving mode as the practical way to be environmental. Fortunately, from this academic perspective, there are plenty of external problems and there is overwhelming evidence that those who support universities would like to see them addressed, or even solved. Nor does this mean that basic or curiosity-driven study would be downgraded or displaced. Perhaps the central environmental problem is that we do not understand the fundamentals of how large systems work; the mechanisms regulating whole river basins, the coupling of forests and climate, and the workings of coastal zones, to name a few, are all still relatively obscure. Nor do we understand how human cultures adapt to changing environments and what human potentials this adaptation releases or suppresses. So there is plenty for everyone to study.

The second problem is to get people, particularly but not exclusively the students, at universities and schools reinterested in science and scholarship in general. This is not a problem specific to the debate in universities about what to do about the environment. It is, rather, a general and alarming product of our current educational society. We are perilously close to becoming a society of the kind called "mandarin," in which qualifications are supreme and the ability to know and do real things is vastly secondary. I trace this directly to the concentration on backward-looking subjects in education, those that depend on views received from the past: law, history (although, paradoxically, history is a discipline with a future), and to a great degree, business, in the case study mode. By no means do I want to imply that these academic pursuits and professions are somehow bad. Indeed, it is their scholarly and worldly successes that have created their intellectual dominance. Rather, it is the weakness of the competition that has left the field and clubhouse to those who primarily interpret events after their time. Science, particularly the practical sciences and science-based professions, from agriculture to zoology, has retreated within itself and has virtually ceased to influence the broader curriculum.

The marvelous empirical hard-nosedness, innovation, and involvement perhaps best exemplified by the old land-grant universities are largely in disrepute or invisible, and with that loss, the real, broad-based interest in science has perished.

The Solution

People are interested in the environment. Universities can use it as an integrating theme. No discipline need be categorically excluded. There are plenty of problems to solve and theoretical and immediate opportunities to pursue. All we need to do is couple in a practical manner the interest in the environment to the teaching of science and scholarly ways. Landfills, like winter wheat and loblolly pine, will yield, literally and figuratively, to hard-nosed empiricism coupled with strong theory. The interest in science thus revived in undergraduates will cause a disciplinary flowering unprecedented in our history as natural scholarly proclivities are matched to real individual and collective interests. The creative energy now under "mandarin" repression will be released, and many of us will be too excited to be lazy or venal.

The Sober Task

Most of us are turned inward on disciplinary, professional, or commodity tracks. We must turn outward: outward from our departments or professions and outward from our institutions. Many will need to think about how to do this. My list of specifics, doubtlessly deficient, follows.

1. The basic course in the basic curriculum should be an environmental problem-solving course treating productivity, sustainable development, environmental ethics, and the application of science and scholarship to problems. All students should take this. It should be taught by the best teachers in the university and should occupy most of everyone's first term.

2. All students should be required to acquire science, history, mathematics, and languages on an individual design derived from their experience in the first term. All undergraduates should design and carry out an environmental scholarship or research project. This will be relatively nonrestrictive, because we have already stipulated that the environment is everything.

3. Subsequent specialization should be allowed, more or less in the mode of current liberal arts majors, but no one should avoid the basics described above.

4. Professional and true disciplinary education is reserved for graduate school; even then, flexibility, communication, and curiosity are inculcated and rewarded.

If these steps are taken, colleges of agriculture will regain their role as shapers of thought and destiny. They will be teaching everyone and not have the sole role of specialized, insular research institutes. They will be doing what they did best in the past: making learning broadly popular and useful. There is a catch. To participate, they will have to lead, and that will require painful and difficult reassessment. But nothing worthwhile, particularly in academia, is free.

References

National Research Council. 1989. Alternative Agriculture. Washington, D.C.: National Academy Press.

National Research Council. 1990. Forestry Research: A Mandate for Change. Washington, D.C.: National Academy Press.

Orr, D. W. 1990. Is conservation education an oxymoron? Conservation Biology 4(2):119–121.

Smith, P. 1990. Killing the Spirit: Higher Education in America. New York: Penguin USA.

Sykes, C. J. 1988. Profscam: Professors and the Demise of Higher Education. Washington, D.C.: Regnery Gateway.

Wilshire, B. 1989. The Moral Collapse of the University: Professionalism, Purity, and Alienation. Albany: State University of New York Press.

6

Environment and Ecology: Greening the Curriculum

A Public Policy Perspective

James R. Moseley

I, like practically every American citizen, am more than willing to say something about education and our educational system. In looking over the information that was sent to me as background for the conference, I read that I should make my presentation from the perspective of someone who is confronted daily with the real-world demands for decisions and policy interpretation relative to environmental issues and what the curriculum through which we educate the food, agriculture, and natural resources professionals for the twenty-first century should provide these young people.

I thought about the assignment and then reflected on some of the things that I had dealt with over the previous week. For example, I became involved in policy related to the free trade agreement between the United States and Mexico (the environment is going to be a key question related to that agreement). I met with lawyers on an Everglades National Park lawsuit. I spent hours on wetlands negotiations. I tried to determine how to keep the courts from shutting down timber production in the Pacific Northwest because of the threat to the spotted owl. I worked on writing the rules and regulations for the 1990 Farm Bill. I prepared to go to Capitol Hill to testify on our 1992 budget. I wrote a briefing for a trip I took to Geneva to establish an international forestry agreement. I worked with the two department chiefs on management strategies for the 60,000 employees in the Soil Conservation Service and the U.S. Forest Service.

My assignment for the conference was to discuss what the curriculum should be to prepare students to address issues such as these. In other words, what courses should I have taken when I

55

was at Purdue University to get myself ready to do the things I listed above?

As I examined this subject, I thought about higher education and realized that there was no curriculum that could ever prepare me to address all of the issues that I must deal with today. When I worked at the U.S. Environmental Protection Agency with William Reilly, the agency's administrator, I wished that I had understood environmental law much better. Now, as an assistant secretary, I wish that I had obtained more experience in forestry, soil science, and public administration. I can make a lengthy list of areas in which I wish I had more experience.

The point is that none of us can predict what the issues are going to be as we move into the next century. As a result, we cannot determine exactly what classes our students should be taking today. I do know, however, that there are some fundamental skills that all of us must master in order to be effective in handling whatever issue comes our way.

In her book *Megaskills* (Rich, 1988), Dorothy Rich defined megaskills as those things that each of us needs in order to succeed. She says they are inner engines of learning and the stuff that achievement is made of. The 10 megaskills that she identified are confidence, motivation, effort, responsibility, initiative, perseverance, caring, teamwork, common sense, and problem solving.

One may think that there is really nothing new to that, and that may be right. These are the basic skills that all of us need to have in order to have a fulfilling life. However, her book focuses on how to teach these skills to our own children as they are growing up. I would like to take it a step farther, though, and suggest that at the fundamental level of preparing and delivering any college course, whether it is a basic introduction to agriculture or organic chemistry, these megaskills should also be reinforced. I am not suggesting, however, that we need to offer specific classes on problem solving, teamwork, or common sense. That is not a good use of anyone's time. I do suggest, however, that as teachers we must make certain that each course that we offer incorporates these key skills.

The skills I need to have in order to handle the daily issues I mentioned above are the foundation on which I can build. When I try to look objectively, then, at young people who are successful and those who are struggling following graduation, the difference generally is not a lack of technical knowledge as much as it is a lack of inter- and intrapersonal skills.

I am reminded of an old Chinese proverb: "Tell me, I will forget. Show me, I will remember. But involve me, and then I will understand." We need to ask ourselves what kinds of teachers and classes we have at the universities: Are they "tell me" teachers and classes, or do they get to the essence of education? That is, do they involve people in the learning process?

When I look back at my college experience, I know I am no different from anyone else. The classes that had the biggest impact and that were the most useful to me are the ones that in some way took me into action and involved me. So the college curriculum should entail the best "involve me" teachers we can find, teachers who incorporate things like cooperative work and study and international travel experiences, mentor programs, and even extracurricular activities.

For some time I have believed, and I continue to talk about, the publish-or-perish mindset that is doing serious damage to our higher educational process. We need a strong research program, and we need some way to report that. We must go back to science; for example, we must go back to science to address environmental issues.

I believe in science and I believe in research, but the publish-or-perish mentality used to evaluate and promote teachers does not necessarily reward those teachers who are the best communicators in the classroom. For several years I have encouraged university administrators to take a serious look at this particular issue and the way we work through this system. Publishing is important, but it should not be the main criterion for an individual's evaluation.

I would also like to see more universities develop partnerships with business and government, partnerships that provide students with firsthand experience on what we mean when we say that school is not a preparation for life, it is life. Learning and education do not end when the sheepskin is placed in our hand; it is only the beginning of lifelong learning.

One of the real pleasures that I have had since I left Purdue has been the opportunity to work with current college students. We live about 20 miles from campus, and over the course of the past 20-some years we have had more than 200 part-time students employed on our farm. And I suspect we have probably had at least 1,000 to 1,200 students come to our farm in various kinds of classes and laboratory settings. Many of them ask me for advice on what they should be doing in school, what classes they should be taking. I always tell them that I cannot make that decision for them. I can tell them, however, some things that I did not do adequately and some things that I would do differently if I had the opportunity to do it over. For example, I would have broadened my curriculum: I would not have specialized quite as much as I did. It is hard to tell an undergraduate not to specialize. They do not listen, because they are focused on getting a job. They think that they must specialize, and they want to make sure that they have adequate skills.

I am not saying, though, that I would not have obtained an agricultural degree. My agricultural degree is very important to me. It is the degree I should have obtained, and I am pleased with what I have achieved. It was critical for me in terms of developing a knowledge base about science, the technology process, and technology transfer. It was also critical because, from my experience in dealing with environmental issues, we simply do not have enough people who understand science and technology.

My career has obviously gone well beyond the fences of the farm. In fact, it has gone well beyond agriculture. The statistics indicate that our current college students' careers will do the same. Therefore, I would have spent more time learning about the social sciences and would have taken courses in history and philosophy, ethics, political science, and law, because my career today primarily focuses around the application of social sciences to the physical science and technology questions that society faces.

I once heard a quote that said, "The man we call a specialist was formerly called a man with a one-track mind." In my job, I cannot afford to have a one-track mind. I must understand the members of the public who want to till the soil, but I also must understand those members of the public who want to leave the soil alone. I must understand the reasons why we should save the spotted owl. But I must also understand the reasons why timber communities are important and vital to the economic well-being of a region.

It is that continual need to balance the needs and the desires of society that requires all of us to think beyond our own small world. As a result, we cannot have focused and specialized minds.

I would also have learned a foreign language. I did not do that, and it was a serious mistake. I would have tried to participate in some kind of foreign-exchange experience as well. There is no question that we have truly become a global village, a global community. I now know with great certainty that learning a foreign language would have been one of the most useful things that I could have done. We can expect knowledge of a foreign language to continue to be important, because the world is going to continue to get smaller as technology, and communications technology in particular, continues to expand.

Finally, I would like to encourage faculty and administrators to be cautious about pursuing what I call the "salad bar approach" to college curricula. There are many people who believe that we should offer a broader range of courses in college. And I suspect that there will be numerous suggestions that we should broaden the curriculum to address environmental issues. As I look at the type of instruction that would have helped me deal with the environmental issues that I must face now, however, most of them would not have come from specific environmental classes. There are natural resource management issues that encompass the ethical, political, social, and scientific views of society. These are at the core of the environmental questions and are essential to helping us find the answers.

Although I wish I had taken a broader range of courses, I do not necessarily mean that I wish that I would have had more electives to choose from. In fact, perhaps I had too many elective opportunities at the undergraduate level. What I would like to have seen was a required core curriculum in the college of agriculture that contained more of the social sciences. At the undergraduate level, we must focus on a more traditional education of the basics of the physical and social sciences. Students of colleges of agriculture should not be able to use a golf class or something similar to obtain graduation credits. That is not the type of course that is necessarily going to help students deal with the issues.

Even though I am now an assistant secretary of agriculture and I am dealing with a multitude of issues at the national level, we should not forget that many of these issues can and should be solved at the local level. We do not need to be spending our time developing curricula specifically so that we develop national or international leaders and secretaries and assistant secretaries of agriculture, because these issues need to be resolved, insofar as possible, at the local level. We need to develop a program that produces people who appreciate the value of a good education; who understand that good education goes beyond just getting a good job after graduation; and who become school board members and county commissioners, 4-H Club leaders, and people who understand what it means to think globally but to act locally. These are the kinds of students who I would like to see graduate from U.S. colleges and universities.

In closing, I want to share part of a letter Abraham Lincoln wrote to his son on his son's first day of school:

World, take my son by the hand. He starts school today. Teach him, but gently, if you can. He will have to learn, I know, that all men are not just and all men are not true. Teach him that for every scoundrel, though, there is a hero and that for every enemy there is a friend. Let him learn early that the bullies are the easiest people in life to beat. Teach him the wonder of books. Teach him that it is far more honorable to fail than to cheat. Teach him to have faith in his own ideas, even when everyone tells him that he is wrong. Try to give my son the strength not to follow the crowd when everyone else is getting on the bandwagon. Teach him to listen to all men, but to filter all he hears on a screen of truth and to take only the good that comes through. Teach him to sell his brawn and his brains to the highest bidder, but to never put a price on his heart and his soul. Teach him to close his ears to the howling mob and to stand and fight when he thinks that he is right. Teach him gently, world, but don't coddle him, because only the test of fire makes fine steel. I know that this is a big order, world, but see what you can do.

That is the essence of what we have to do.

Reference

Rich, D. 1988. Megaskills: How Families Help Children Succeed in School and Beyond. Boston: Houghton Mifflin.

The Inherent Value of
the College Core Curriculum

Lynne V. Cheney

Through the centuries, the study of subjects like history, literature, and philosophy has brought great satisfaction to people. St. Augustine once said that the only reason to study philosophy was "in order to be happy." A twentieth century philosopher, Charles Frankel, explained the joy that the humanities can bring by noting that people's experiences are enriched if they know the background of what is happening to them, "if they can place what they are doing in a deeper and broader context, if they have the metaphors and symbols that can give experience a shape." Frankel himself used a metaphor to make the point: "Think of what the lore and legend, the analyses and arguments, that surround baseball contribute to our enjoyment of the game. They *make* the game, as anyone can discover by sitting next to someone who is uninitiated" (Frankel, 1981:9–10). The humanities, he argued, with myth, story, and debate, initiate us into life.

The humanities are valuable to us not only as individuals but also as a polity. Knowledge of the ideas that have molded us and of the ideals that have mattered to us functions as a kind of civic glue. Our history and our literature give us symbols to share and to help us all, no matter how diverse our background, to feel that we are part of a common undertaking.

There is a story that illustrates this well. It comes from the autobiography of a woman named Mary Antin (Antin, 1969), who emigrated from Russia to the United States in the early part of the twentieth century. In *The Promised Land*, she writes about becoming acculturated in the United States and going to school. She remembered one day she went to school and was feeling, as many children do, not very important, not very noticed, not very honored or valued by her peers. She learned about a person called George Washington, a man who was revered by his contemporaries and who was honored above all others in his time; she wrote, "I discovered that I was more nobly related than I had ever supposed. . . . George Washington . . . was like a king in greatness, and he and I were Fellow Citizens" (Antin, 1969:224).

Communicating to the next generation the figures, ideas, and events of the past is, for many reasons, a deeply important task. How well are we accomplishing it? Not very well, and not as well as we should, according to a Gallup survey of college seniors that the National Endowment for the Humanities funded a few years ago (The Gallup Organization, 1989). In that survey, 25 percent of the nation's college seniors were unable to locate Columbus's voyage within the correct half-century. About the same percentage could not distinguish Churchill's words from Stalin's, or Karl Marx's thoughts from the words of the U.S. Constitution. The survey gave a list of phrases and some sentences and asked students to answer, true or false, whether the phrases are in the U.S. Constitution. One of them was Karl Marx's phrase, "From each according to his ability, to each according to his need." One of four college seniors thought it was in the U.S. Constitution. More than 40 percent could not identify when the Civil War occurred. The majority could not identify Magna Carta, the Missouri Compromise, or Reconstruction. Most could not link major works by Plato, Dante, Shakespeare, and Milton with their authors. To the majority of college seniors, Jane Austen's *Pride and Prejudice*, Dostoyevski's *Crime and Punishment*, and Martin Luther King, Jr.'s "Letter from Birmingham Jail" were clearly unfamiliar.

It is not just in the humanities that college seniors are found wanting. The National Science Foundation recently sent a film crew to a Harvard University graduation and interviewed the bright, fresh-faced students who were about to receive their bachelor's degrees. They were asked to explain why there are seasons. All of the graduates answered with an air of great authority, and also with complete inaccuracy. Most of them explained that there is winter because the earth is farther from the sun then. Even if you do not know the right answer to this question, you could quickly figure out that the answer the college graduates gave was not correct. If the essential point is that the earth is farther from the sun, then why is it not winter everywhere, in Canberra, Australia, as well as Cambridge, Massachusetts?

So it is important to note that college seniors' lack of basic knowledge of major areas of human thought is not limited to the humanities, and it is important to remember that the failings described by these examples are not simply the result of 4 years of higher education. It is the result of at least 16 years of schooling. Nevertheless, it is possible to look at our nation's college campuses and see an important part of the reason why we have college seniors who cannot tell Churchill's words from Stalin's or a good scientific explanation from a bad one.

Students can graduate from 80 percent of the nation's 4-year colleges and universities without taking a course in the history of Western civilization. They can earn a bachelor's degree from 38 percent of U.S. colleges and universities without taking a course in history, from 45 percent without taking a course in American or English literature, from 41 percent without studying mathematics, and from 33 percent without studying natural and physical sciences.

In the report that we issued along with our Gallup survey, *Fifty Hours* (National Endowment for the Humanities, 1989), the National Endowment for the Humanities recommended a required course of study, a core of learning, to ensure that undergraduates have opportunities to explore in broad-ranging, ordered, and coherent ways the major fields of human inquiry. This core, we suggested, might make up about 40 percent of the curriculum, or 50 hours, a figure that falls well within the range of general education credits required by most colleges and universities.

As it is now, however, the general education credits that could be devoted to a core curriculum are all too often organized into loosely stated distribution requirements. These requirements mandate that students take some courses in certain areas and some in others, and in catalogues there are very long lists of the acceptable choices. For the most part, they are specialized offerings. They often have very little to do with the broadly conceived learning that should be at the heart of general education. Indeed, some of the courses that seem to relate to, or that are listed as being part of, certain areas of knowledge seem to have little to do with the areas of knowledge that they are supposed to elucidate.

At one public university in the West, it is possible to fulfill humanities requirements with courses in interior design. In 1988-1989, at a private university in the East, one could fulfill social science distribution requirements by taking lifetime fitness. At a midwestern university, students can choose from almost 900 courses, with topics ranging from the history of foreign labor movements to the analysis of daytime soap operas.

The result of this kind of narrow presentation of giving students long lists of courses that are very narrowly conceived is what naturalist Loren C. Eiseley relates to what he once described as a meaningless mosaic of fragments. In his book *The Unexpected Universe*, he writes, "From ape skull to Mayan temple, we contemplate the miscellaneous debris of time like sightseers to whom these mighty fragments, fallen gateways, and sunken galleys convey no present instruction" (Eiseley, 1964:6). It is a wonderfully written phrase that describes all too well the state of the curricula on many college campuses.

A core of learning, on the other hand, can show a pattern to the mosaic. Taking what John Henry Cardinal Newman once called a "connected view of the old and new, past and present, far and near" (Newman, 1982:101), a core can provide a context for forming the parts of an education into a whole.

In *Fifty Hours*, we set out one scheme for a core curriculum. We also examined the colleges and universities that have succeeded in establishing core curricula and found that it is being done in every part of the country. Although the number of these colleges and universities is still relatively small, their numbers are growing. The pace of change is still slow, though, and it is no doubt in part because the task of designing a rigorous and coherent plan of study is hard, as we discovered when we did it at the Endowment. It undertakes to answer a very challenging question: What should an educated person know?

It is challenging professional groups as well. As I have talked about the subject of core curricula across the country and have had faculty members tell me about their struggles with establishing a core curriculum on various campuses, I have been impressed by the departments and disciplines they indicate are the obstacles to establishing a core. It is not colleges of agriculture but colleges of engineering that demand so much of the curriculum that it is very difficult to find time for general education. The other culprit named most often is music educators, who also demand a great deal of the core curriculum for future music teachers and who make the establishment of a core curriculum for general education difficult.

There are other obstacles to establishing a core curriculum. One is that we have found an intellectually respectable way to argue that we do not need one. We say that it does not matter so much what a person knows in various fields; we have started to say that what really matters is that a person understands the methods of inquiry that are used in different fields. The issue is not knowledge but "approaches to knowledge," which is a quote from Harvard's catalogue.

There has been debate about the Harvard core curriculum since its beginning. It has been covered in the popular press as well as in the academic press. Critics argue that no matter how good the courses in what Harvard calls its core curriculum are and no matter how fine the faculty members who teach these courses are, taken together they do not provide the connected learning discussed by Cardinal Newman. The courses are a miscellaneous assemblage, critics say, rather than a coherent framework for learning. For example, a student can satisfy the history requirement by studying the Cuban Revolution or tuberculosis in the nineteenth and twentieth centuries. A student can satisfy the literature requirement by studying either Shakespeare's later plays or, as stated in the catalogue, "beast literature."

There are many reasons to explain why these narrow courses are offered. One is the emphasis on research rather than teaching that we have seen on many college campuses. The fact that all rewards go to research and few go to teaching means, on the one hand, that there is very little reward for people who want to do the hard work of getting a core curriculum together. On the other hand, it means that without these broadly conceived courses being put into place, people will teach their research interests. That is what accounts for a course on nineteenth and twentieth century tuberculosis in the core curriculum at Harvard: It may be an important and interesting research topic, but it does not make much sense as a requirement for an undergraduate in general education.

To be fair, there is another side to the argument. That argument is that Harvard does not define intellectual breadth as the mastery of a set of great books, the digestion of a specific quantum of information, or a survey of current knowledge in certain fields. Instead, Harvard seeks to introduce students to the major approaches to knowledge in areas that faculty consider indispensable to undergraduate education.

Harvard is not alone in this allegiance. It can be found in many institutions of higher education, and it can be found in primary and secondary schools as well. It is important to remember that these things are a continuum, that what happens in colleges and universities almost inevitably trickles down to primary and secondary schools, where an emphasis on ways of knowing and on the process of knowing rather than on knowledge itself permeates almost every grade. In the various early grades, basal readers are used to teach students how to read. Again, their approach is the one that Harvard recommends. At the earliest level, students are not given real literature or introduced to real stories; rather, they are given approaches to knowledge, ways of thinking.

The mental skills that these books aim to teach are often listed in the front of teachers' editions. The mental skills they attempt to develop usually involve such things as how to identify the sequential order of events or how to follow directions involving substeps. The basal readers try to teach these things by constructing a plastic, artificial prose, not real stories.

One mental skill that the basal readers particularly emphasize is how to find the main idea, which is an aptitude we all want our children to have. In looking through many basal readers, however, I have come across pages on which children have been instructed to find the main idea and have discovered that there was absolutely no main idea worth finding. This exemplifies the difficulty of trying to teach skills without paying sufficient attention to content.

Another extreme manifestation of this syndrome can be found at education conventions, where publishers fill their display racks with row after row of books that promise to teach youngsters how to think. These books are not quite content-free, but they come as close as possible. Their mainstays are exercises in seeing analogies: Is a triangle more like a human being or a wheelbarrow? Meanwhile, looming over the educational landscape is the Scholastic Aptitude Test (SAT), an examination that, in its verbal component, studiously avoids assessment of substantive knowledge. The SAT is indifferent to whether test-takers have studied the Civil War, learned about Magna Carta, or read *Macbeth*.

The emphasis that the SAT puts on what its creators call *developed ability* as opposed to knowledge makes this test unique among those used in industrialized nations. While our students are trying to sort out antonyms and analogies on the SAT, students in Japan are writing about the foreign policy of Afghanistan. Students in Germany are writing about the Weimar Republic and the development of democracy there. Students in France are writing about U.S. foreign policy from Jimmy Carter to George Bush. While students abroad are being tested on what they have learned, we are testing developed ability, which has nothing to do with what students have learned in the classroom.

Another example of the way in which, throughout our educational system, we have elevated the process, the ways of knowing, and the ways of thinking above knowledge itself is in the field of English. The approach is called *discourse studies*, which is an approach to knowledge that has become enormously influential in literature. What counts most in such teaching and research is not the "what." The subject can be anything: a poem, a play, or a bumper sticker. What counts is the "how": How is this text, seemingly innocent, implicated in ideology? That is the question that is asked. How can it be unmasked?

At the University of Minnesota, the Humanities Department recently proposed abolishing its chronologically ordered Western civilization sequences and substituting three new courses: the first is called Discourse and Society; the second is called Text and Context; and the third is called Knowledge, Persuasion, and Power (University of Minnesota, 1989).

In the old courses, the focus was on the works of Plato, Dante, Descartes, and Rousseau. In the new ones, the emphasis is on "the ways that certain bodies of discourse come to cohere, to exercise persuasive power and to be regarded as authoritative while others are marginalized, ignored, or denigrated." Instead of focusing on the writings of Wordsworth or Eliot, the new courses emphasize "hegemony and counterhegemony."

Given the pervasiveness—or hegemony—of the view that ways of knowing should have preeminence over knowledge, the time has come for a thoughtful and thorough examination of this idea.

Let me pose two questions. First, even if we posit that the various fields of human inquiry are at the very highest levels of scholarship and are distinguished by differing approaches, is this a matter of any interest or use to undergraduates? I approach this question from the standpoint of literature, and I must say that most undergraduates and people I have known who love novels, plays, and poetry are not interested in them as methods of discourse but as sources of insight into their lives and into the human predicament. Author Annie Dillard once asked, "Why are we reading, why would we read, if not in hope of beauty laid bare, life heightened, and its deepest mystery probed?" (Dillard, 1989:1). There is satisfaction, of course, in seeing how language achieves beauty, heightening, and revelation. It is the achievement itself, however, the novel or book or play, that draws most people back time and again.

My second question is this: When throughout our system of education we emphasize approaches to knowledge, what kind of young people are we likely to produce? Even if we assume that it is possible to teach the ways and processes of knowing without emphasizing knowledge itself, then we could hypothesize a quick-witted, nimble-brained generation that, perhaps not knowing as much as they should, nevertheless has the ability to learn quickly. It may also be the case, however—and this is the point that E. D. Hirsch (1987) makes—that not knowing as much as one should severely hinders one's ability to know and to learn more, much less to learn quickly.

Bernard Lewis, who is Princeton's distinguished professor of Islamic studies, recently told of teaching a graduate seminar and finding that the students in it did not know what the Crusades were. They had the modern meaning; they knew about a crusade as a cause. But they had no idea of the word's historical significance. This would be a great hindrance to students engaged in advanced study of Middle Eastern history.

Lack of knowledge can be an obstacle to understanding the present as well as the past. In 1989, a story in *The Washington Post*, "Teenagers Find East European Events Confusing—or Irrelevant" (Maraniss and Peterson, 1989), told of teachers across the United States trying to engage their students with the dynamic and moving events occurring in Poland, Hungary, Germany, and Czechoslovakia and of those teachers finding their students confused and indifferent. The students simply did not have a sufficient historical context to understand the significance of the changes. As one teacher put it, "They don't understand what communism is in the first place. So, when you say it's the death of communism, they don't know what you're talking about." During a discussion in which the former Eastern bloc countries were referred to as "satellites" of the Soviet Union, one student raised her hand and said, "I'm sorry, but what is this talk of satellites? Are we talking about satellite dishes or what?"

In our educational system, the emphasis on approaches to knowledge as opposed to knowledge itself is not the only culprit. We can find many reasons why students do not know as much as they should. But, surely, the emphasis on process and the neglect of content seen at all levels of education are important factors. If we do not emphasize that there are some figures, books, and events that are important to know, then we should not be surprised when young people do not know them. And if we do not undertake the hard work of setting out a framework for learning, then we should not be surprised when students do not have one and when they have difficulty making sense of new events.

Concentrating on knowledge, on what should be taught and learned, as well as on ways of teaching, learning, and knowing, is not easy work. It may, however, be among the most important efforts that those of us who care about education can undertake.

References

Antin, M. 1969. The Promised Land, 2nd ed. Boston: Houghton Mifflin.

Dillard, A. 1989. Wait till you drop. The New York Times Book Review, May 28, 1989, p. 1.

Eiseley, L. 1964. The Unexpected Universe. New York: Harcourt, Brace & World.

Frankel, C. 1981. Why the Humanities? Address delivered at the Lyndon Baines Johnson Library, University of Texas, Austin, December 4, 1978. (Published in National Humanities Center. 1981. *The Humanist as Citizen*. Chapel Hill: University of North Carolina Press.)

The Gallup Organization. 1989. A Survey of College Seniors: Knowledge of History and Literature, conducted for the National Endowment for the Humanities. Princeton, N.J.: The Gallup Organization.

Hirsch, E. D. 1987. Cultural Literacy. Boston: Houghton Mifflin.

Maraniss, D., and B. Peterson. 1989. U.S. students left flat by sweep of history; Teenagers find East European events confusing—or irrelevant. The Washington Post, December 2, 1989, p. A1.

National Endowment for the Humanities. 1989. Fifty Hours: A Core Curriculum for College Students. Washington, D.C.: National Endowment for the Humanities.

Newman, J. H. 1982. The Idea of a University. Notre Dame, Ind.: University of Notre Dame Press.

University of Minnesota. 1989. Humanities Department curriculum. University of Minnesota, Minneapolis. Photocopy.

8

General Education and
the New Curriculum

Gary E. Miller

Historically, concern over the undergraduate curriculum seems to coincide with times of major change or disruption in society. The times are now changing once again. What is frustrating is the nature of the change. It is not as if a revolution has overthrown all of the old knowledge, leaving us to teach brand new subjects. It is not as if we have discovered something so new that we have to go back and rethink everything we used to hold as true. It has not been that kind of change. Instead, it seems that much of the old knowledge is still valid, except that the context has changed. It is a difficult and frustrating problem. It does not necessarily require that we throw away everything we are used to doing, but that we revisit how we define what we do.

This is one of the problems with discussing the curriculum. Traditionally, we have tended to treat rather loosely some of the basic tools of our trade. We use terms such as *core curriculum, distribution system, liberal education*, and *general education* as if they are interchangeable. Much of our problem with the revitalization of undergraduate education lies in our habitual use of these terms. Our success or failure in reforming undergraduate education will depend, in larger measure than we usually are willing to recognize, on our ability to revisit and redefine the basic language of the curriculum, so that our actions better reflect our rhetoric.

Many people who read this volume have had the experience of working on a curriculum committee. We discuss goals and make decisions about the curriculum. Then, as we return to our departments, things change. Our interpretation of the goals begins to drift from what we thought were commonly held assumptions. Somehow, what happens in the classroom—in the interaction between the faculty member and the student, where the curriculum really resides—does not reflect the goals we set down on paper. More disturbing, there is no coherence among courses. Yet, each faculty member might feel confident that he or she is living up to the goals.

I begin by looking at some of the terms we use and what they mean. We use some terms as if they define the curriculum, although they do not; instead, they define the structure on which a curriculum can be built. Terms like *core curriculum, distribution system, learning contracts,* and even *interdisciplinarity* reflect the way in which the curriculum is organized rather than the content and goals of the curriculum. They are simply the medium of the curriculum. The caution, then, is not to confuse the medium with the message.

However, our choice of medium does have something to say about the message—in this case, about the assumptions we bring to the curriculum. A distribution system suggests that although we want our students to gain familiarity with a broad range of knowledge—and perhaps some depth of knowledge in a particular area—we are not concerned with the specific knowledge that they get. A core curriculum, on the other hand, makes clear that we hold specific areas of knowledge to be very important, so important that all students should share them, regardless of their personal or professional interests. Interdisciplinarity—which can function in either a distribution or a core curriculum—suggests that we recognize that the way in which we compartmentalize knowledge for research and organizational purposes is not necessarily relevant for instructional purposes. All three of these tend to assume that knowledge is the central force in the curriculum. By comparison, some structures—experiential learning and learning contracts, for example—assume that the individual personal and professional interests of the student come first.

The interplay between the purpose of a curriculum and how we structure the curriculum is obvious. We should not assume that structure is independent of purpose or that any one structure is the universally appropriate one. That decision should rest on how we define the goals and our approach to the content and methods of the curriculum.

General education and liberal education are two of the most common ways that we define the purpose, content, and methods of the curriculum. Just as we use terms like *core curriculum* and *distribution* interchangeably, we have used terms like *general education* and *liberal education* interchangeably, even though they arise from distinctly different roots. In fact, I read one article in which the writer alternated the use of "liberal/general education" with "general/liberal education" in an attempt to avoid making a distinction. They are not the same, however. They reflect fundamentally different assumptions about the purpose of education, about the nature of knowledge, and about the relationship of the learner to the curriculum. Because these distinctions are absolutely vital—and not necessarily apparent—I would first like to define general education and then compare general and liberal education.

General education is a curriculum movement that grew out of the philosophy of American pragmatism in the early part of the twentieth century. Initially, it was a response to the increasing "professionalization" of the liberal arts (which themselves had grown rapidly toward the end of the nineteenth century as a response to the fragmentation that accompanied the growth of research as a mission of the university). However, general education soon took on a meaning separate and distinct from liberal education. I define it in this way:

> General education is a comprehensive, self-consciously developed and maintained program that develops in individual students the attitude of inquiry, the skills of problem solving, the individual and community values associated with a democratic society, and the knowledge needed to apply these attitudes, skills, and values so that the students may maintain the learning process over a lifetime and function as self-fulfilled individuals and as full participants in a society committed to change through democratic processes. As such, it is marked by its comprehensive scope, by its emphasis on specific and real problems and issues of immediate concern to students and society, by its concern with the needs of the future, and by the application of democratic principles in the methods and procedures of education as well as the goals of education (Miller, 1988:2).

This definition contains several key phrases that deserve closer scrutiny. First, general education is "comprehensive." By that I do not mean that it covers the entire canon of Western civilization. Instead, general education is comprehensive in that it deals with basic contexts, methods, attitudes, values, and skills that apply in all areas of our students' lives. It is also comprehensive in that it applies to the entire learning environment. It is not limited to the first 2 years of a program; it ideally is integrated with the total curriculum, including the upper-division professional programs and those areas that we do not tend to associate directly with the curriculum, such as faculty development, student participation in the service function, and sponsored extracurricular activities.

Second, general education is self-consciously developed and maintained. The stated purposes of the curriculum guide every aspect of the curriculum, including the selection of subject matter, the structure, and the methods and procedures. Evaluation is done continually, with an eye to improving the match among goals, methods, and outcomes.

General education is concerned with the individual student. It is a student-centered curriculum. It is concerned explicitly with the development of the learner rather than with the delivery of instruction. This is not necessarily an easy distinction to draw, but it is a crucial one.

General education is intimately concerned with democratic processes and with the needs of a democratic society. It begins with the individual and his or her relationship to society as its first organizing principle. Its goal is to enable individuals to perform their basic responsibilities as members of a democratic society. It assumes that society is dynamic and that the ability of the individual to participate in setting the direction of change is fundamental to the health of a democracy. This suggests another characteristic of general education: it is concerned with specific, immediate issues and with the present and future rather than with the past.

In all of this, there is another characteristic that is central to a general education and that has a significant effect on how general education curricula take form. That is that the means must support the goals of the curriculum. Here, the structure and content of the curriculum and the ways in which students interact with subject matter, faculty, and their environment come together.

General and Liberal Education

General education shares many of its goals with modern liberal education. However, the two build on different assumptions that affect the content and methods of the curriculum. These differences in basic assumptions and goals are very important if we want to ensure that the goals are realized in the classroom.

General and liberal education have fundamentally different assumptions about the role of knowledge in education. Liberal education, building on the classical liberal arts, assumes that knowledge is valuable in its own right. General education, on the other hand, grows out of U.S. pragmatism. Its explicit goal is to equip students with the skills they need to control their environment. For that goal, knowledge is a tool rather than an end in itself.

Traditionally, liberal education is concerned with ideas in the abstract, the preservation of universal truths, and the development of the intellect. General education is concerned with the development of the student's capability for individual and social action and with the problems of the present and future.

If these differences sound theoretical, they become very real in the classroom. To illustrate, look at how a typical course might be treated differently in three situations—a simple distribution approach, a liberal education approach, and a general education approach. Take Introduction to Art History. In a traditional distribution curriculum, Introduction to Art History would be just that—an introduction to the study of art history. We would learn about the development

71

of various techniques and schools of art—the development of perspective, the growth of realism, the discovery of light as a tool, the evolution of abstractionism, etc. In a liberal arts curriculum, we would be more likely to study the history of art as an expression of culture. We would study the evolution of art as religious expression, the return to Classical motifs in the Renaissance, and the gradual emergence of art as an expression of the individual and of social concerns. In a general education curriculum, Introduction to Art History would focus on art as an example of how people in a particular time and place perceived their world and used art to respond to their environment; students might be asked to examine social trends that parallel changes in the style or content of painting across several periods, for instance.

The same differences in goals, treatment of content, and methods would apply as the different approaches were used in professional education curricula. The key is the interplay among underlying assumptions about the purpose of the curriculum, the goals that grow out of those assumptions, the way content is selected and treated to meet those goals, and the methods and structure that are brought to bear.

Why General Education?

Given the choices, why choose general education? The first question is this: Why change what we are already doing at all? The fact that people attended the conference on which this volume is based suggests that we sense the need for change. But why? Is it because something in the curriculum is broken and needs to be fixed? Or is it because the job itself has changed and the old tools are no longer right for the job? I submit that it is the latter. It is not that we have forgotten how to teach or have wandered away from the path. It is that our world has changed—is still changing—and with that the context for teaching and learning has shifted. As a result, we must teach some new things, teach some old things in a new context so that they are more useful in the new environment, and find some methods and structures to make the curriculum coherent.

In order to know how to adapt, we must know the direction of change in society. In a recent article entitled "The Real Economy," Robert Reich suggests that direction from the perspective of an economist. He maintains that, in the new world economy in which we find ourselves, employers need several core skills. These include, "the *problem-solving* skills required to put things together in unique ways (be they alloys, molecules, semiconductor chips, software codes, movie scripts, or pension portfolios), . . . *problem-identifying* skills required to help customers understand their needs and how those needs can best be met by customized products, . . . the skills needed to link problem-solvers and problem-identifiers," what Reich calls "strategic brokers" (Reich, 1991:37).

It seems that the assumptions that underpin general education are very much in line with the direction of social change, at least as people like Reich see it. We have a match of social need and curriculum goals. Moreover, with the technologies that are available to us, we have at hand the educational means to realize the objective.

Implications for Agriculture-Related Curricula

The first thing to remember is that general education is not separate from professional education. Instead, it is a way of defining the mission and goals of the total curriculum. The purpose of general education is to develop the student's ability to function effectively in society—as an individual, as a family and community member, and as a professional. The characteristics of general education that I just described should infuse the entire curriculum, so that we produce professionals who are able to see their professional work within the context of their broader community responsibilities and, more importantly, who are able to act on that vision.

General education's goals of problem solving, decision making, and values clarification and its expressed commitment to immediate issues and enabling students to shape their future suggest methods well-suited to the experiential, practice-research orientation of agriculture and natural resources-related curricula. Inquiry and experimentation are central methods of a general education curriculum.

Some other changes in method should also be considered. Although traditional curricula focus on individual competitiveness and achievement, a general education curriculum might also include group problem solving—a technique that more accurately reflects how we work in daily life and that develops the student's ability to be effective in a variety of communities with a variety of constituencies.

Beyond the professional curriculum, I believe each profession has a right to demand a relevant preparatory curriculum for students. By this, I do not mean the Shakespeare for Engineers courses taught at many institutions. However, an American history course that uses history of land use rather than political history as the connecting theme might be more relevant not only to students in colleges of agriculture but to others as well. A course in humanities that explores the evolution of perceptions of human interaction with the natural world could be just as effective—possibly more effective—than a traditional history of philosophy or introduction to the humanities course. The common thread in these courses is the distinction between learning something and learning about something. In focusing on the former, they increase the chances that education will lead to lifelong interest in these dimensions of life.

One way to encourage general education is to work in faculty teams to develop course materials and methods that can then be shared. This approach has several advantages. First, it encourages interdisciplinarity. Second, it allows for peer review during the course development process. Third, it frees courses from the need for lectures, allowing more time for competency-based learning and a more fulfilling role for faculty.

Until recently, we emphasized knowledge and theory because it was difficult to give students direct experience. Today's technologies allow for computer simulations, videotapes that support inquiry by allowing students to observe a shared experience, and interactive video modules that allow students to safely experiment with, for example, land use decisions. These technologies are just now beginning to find a place in higher education. They hold particular potential for helping us to realize the goals of general education by providing opportunities for experience, to simulate the impact of decisions over time, or to see the consequences of what would otherwise be dangerous, if not disastrous, action.

Finally, a general education program will be successful only to the extent that both the students and the faculty understand its objectives and participate in it self-consciously. Curriculum development must go hand in hand with faculty development. This could include ongoing faculty seminars on the major issues to which the curriculum is addressed. Or it might mean creating new relationships between faculty and students by treating faculty research issues as cases for instructional inquiry, bringing students and faculty together around real issues of immediate concern to both.

Conclusion

General education is not a single curriculum. It is an approach to building and maintaining a curriculum. The final curriculum should reflect the mission of the institution, its location, and its traditions. A general education curriculum can include components of liberal education; it can be highly structured or student initiated; it can be a lot of different things. But all general education curricula share the basic characteristics that I have discussed here. The key lies in taking seriously the issues of defining our terms, of testing—and constantly retesting—our assumptions, and by keeping clearly in sight the goal of preparing students to live and function as professionals in their future.

References

Miller, G. E. 1988. The Meaning of General Education: The Emergence of a Curriculum Paradigm. New York: Teachers College Press.
Reich, R. B. 1991. The real economy. The Atlantic 267(2):35–52.

9

Agriculture: A System, a Science, or a Commodity?

Norman R. Scott and Brian F. Chabot

Although past agricultural policy focused on production and farm commodities, it is clear that future agricultural policy will be driven by environmental issues, rural development, food safety, energy, information technologies, and global competition. This chapter has two basic objectives. The first is to ask the following question: What is agriculture and what must colleges of agriculture become, especially relative to undergraduate education through the 1990s and into the twenty-first century? The second, more significant objective is to show that we must place much greater emphasis on our perceptions of the colleges of agriculture of the future.

What is Agriculture?

In the beginning, agriculture simply meant farming. Agriculture was a word that "educated people" used to refer to farming. The food distribution system was rudimentary, value-added processing occurred in the home, and farm implements were produced by a blacksmith rather than an agricultural implement industry.

Agriculture is defined as the art or science of cultivating the land; the production of crops and livestock on a farm; farming. As we view agriculture today, we see it as a system of farmers and agribusinesses that supply production resources (machinery, fertilizer, money, etc.) or that process and distribute products from the farm. Consumers, as individuals who purchase agricultural products, thereby influence what is produced and are an integral part of the system. Community services surrounding farm families and agricultural industries are a key part of the system.

The public's consciousness and concern for the environment has

75

moved it to one of the higher national priorities. The public's concerns about food safety with respect to pesticides; a growing concern for the quality of groundwater and surface water resources; increasing attention to global climate change; and the issues of economics, energy, and biotechnology illustrate the heightened public consciousness. Environmentalists and animal rights activists are vocal parts of the system. In the cover story of the September 26, 1988, issue of *Fortune* magazine, which focused on managing for the 1990s, the following question was posed: What issue will grab people the most? "Despite mounting distress about AIDS, drug abuse, and the homeless, some observers think that the number one issue will be environmental protection. We will be obsessed with water in the 90's" (Kupfer, 1988:45).

Building upon the statement of John Muir, that "everything is hitched to everything else in the universe" (Muir, 1911:53), the boundaries of any system are more difficult to define. The point is that the agricultural system is much larger than it used to be and increasingly is more tightly interwoven into the full fabric of society. Thus, for the purposes of this chapter, we define agriculture as the activity of mankind that produces healthful and nutritious foods, industrial feedstocks, and renewable fuels while enhancing and maintaining the quality of the environment; energy, raw materials, and food are undeniably necessary for a stable society.

Roles of Colleges of Agriculture

Because colleges of agriculture played a lead role in creating the complex and diverse food and agricultural system that we have today, it is not surprising that colleges of agriculture today are far different from their humble beginnings approximately 125 years ago. At their inception, a mere handful of faculty focused their efforts on training young men in the practical arts and emerging science of farming. Research to solve the problems of the farm soon followed. Later still, an active extension responsibility was added to the duties of the faculty. This evolution of function paralleled an evolution of the kinds of courses that were offered and the subjects that were researched. The early history of Cornell University, which probably parallels that of other land-grant institutions, documents a struggle to find individuals capable of teaching agricultural subjects. Those hired as faculty tended to be trained in the classical subjects of chemistry and biology, but with a bent toward agriculture. Practical agricultural training was initially handled by successful local farmers interested in the future of the new college—a reversal of today's extension process. Today's agriculture faculty at Cornell University comprises more than 450 people (including many from the College of Agriculture and Life Sciences, College of Veterinary Medicine, and College of Human Ecology). The faculty's expertise captures the complexity of contemporary agricultural systems and rural issues, with a range of course offerings so vast as to be inconceivable to those early pioneers on a fledgling faculty.

Through this process, universities in the United States evolved away from providing a limited classical education for a privileged few to serving the aspirations and needs of society as a whole. Certain of these needs are of paramount importance because they deal with the fundamental resources upon which our civilization is built. Primary resource areas are food, energy, environment, and economics. Many other program areas are crucially important to the educational and research roles that universities play. However, those universities that aspire to serve national needs must have well-developed programs in most or all of these fundamental resource areas. It is within this larger context that our vision for the role of colleges of agriculture is presented.

In order to gain some perspective as to how others view the role of colleges of agriculture, we did a quick survey of deans and directors of Cornell's College of Agriculture and Life Sciences, asking them what they regarded as the mission of the college. Here are the replies:

• to conduct programs in research, extension, and instruction to meet the needs of the people of New York State as a land-grant college;

• to generate knowledge and transmit that knowledge to clients in the state, nation, and world and to produce young men and women for leadership positions;

• research and education;

• to educate students, to create new knowledge, and to disseminate knowledge to various publics;

• to be the number one land-grant college; and

• to support (sustain) the agricultural industry of New York and the people it serves.

As can be seen from these responses, present leaders of Cornell's College of Agriculture and Life Sciences tend to view its mission in broad terms. It is not until priority programs are defined that the college begins to distinguish itself from other colleges at the university. This larger world view has developed with time, because the perceived mission of the college at its founding was focused on agriculture in a less ambiguous way.

Given that the agricultural system has grown more complex, what, then, do we view as the role of colleges of agriculture?

First, it needs to be emphasized that training of undergraduates is only one of the roles of colleges of agriculture. Even if there were no undergraduate students, research and extension needs would argue for the continued existence of faculty in colleges of agriculture. Colleges of agriculture can expect to continue to provide the research base for farmers, packers, shippers, wholesalers, retailers, bankers, rural schools, community planners, and the many other professions that relate to modern agriculture.

Second, colleges of agriculture should draw on the wealth of scientific expertise to address the more general issues now facing society. For example, expertise in farm finance is easily extended in the curriculum as accounting, business management, and private entrepreneurship. A program in communications for extension professionals can produce courses in scientific writing, speaking, video production, and more. Basic biological sciences faculty in production agriculture departments can contribute to a general biology curriculum for the university as a whole. Faculty with expertise in water-quality management or energy use on the farm can contribute to larger programs on environmental issues or energy policy. This line of thinking extends to many other areas. Faculty in colleges of agriculture have special expertise that can relate to the more general interests of the current students. Stability of enrollment within Cornell's College of Agriculture and Life Sciences has come entirely through reframing the roles of individual faculty and groups of faculty.

Along these lines, colleges of agriculture have much to contribute to what we suggest are the fundamental and enduring resource issues faced by our society: food, energy, environment, and economics. Drawing on a long history of involvement and a significant depth of expertise, colleges of agriculture will be among the strongest players in these areas. Some colleges have already made this transition.

Third, colleges of agriculture should accept the responsibility for providing a general education in agriculture for nonagricultural students. Gould Coleman, an historian at Cornell, relates the following about George Stanton Gould, a charter member of Cornell's faculty. Gould viewed agriculture as a framework within which a vast part of man's knowledge could be fitted. He believed that no man could be educated without some exposure to agriculture, and from this perspective, he gave a series of lectures on agriculture at large to the entire senior class of the university. The inspiration for these views has been lost in the current curriculum. We do not have such a course today, but there is not a better time or a greater need for one. Beyond this, who is better positioned to teach courses in issues of food safety, nutrition, and the environment? This provides a base for what we contend to be a role for colleges of agriculture in the future.

These suggestions presuppose that the need to educate students to be farmers will continue to diminish. We simply have become too good at increasing the production efficiency of farms. Some claim that agricultural research, which is done principally at the land-grant universities, has put farmers out of business. This is a major issue in the debate over recombinant bovine somatotropin. Whatever the cause, it seems inescapable that there will be fewer farmers in the future. In response, the teaching capacity of departments of production agriculture needs to convert to other functions, as described above, or be directed at new audiences.

However, this point raises an issue that must be confronted: Are colleges of agriculture, currently or as they are likely to be in the future, attractive places for training future farmers? At a place like Cornell, with its Ivy League image, it is hard to imagine that peer pressure and the structure of the curriculum are really conducive to farm youth who wish to remain in farming. The most successful farmers in the future are likely to be skilled businessmen with advanced educations and technical knowledge that colleges of agriculture are in the best position to provide. Especially in the face of increasing opportunities and the need to expand into nonagricultural areas, we strongly urge these colleges to examine carefully the kind of experience and educational environment being offered to farm youth.

Can Undergraduate Enrollments be Enhanced in Colleges of Agriculture?

We suggest that there are at least three ways to enhance undergraduate enrollments in colleges of agriculture. First, expand the range of course offerings in colleges of agriculture, especially in those disciplines that are part of the agricultural system but that are pertinent to other elements of society, such as business management; personal enterprise; communications; engineering; and the biological, biomedical, and environmental sciences. Other options include general agriculture and social issues courses dealing with, for example, nutrition, food safety, and health. Many colleges have already moved in this direction, some so substantially that agriculture is now a minor theme.

The curriculum must address science-based agriculture. It must recognize the entirety of the system and its complex interactions. It must be exciting and relevant to the interests of society. It must educate teachers, focus on the basics, identify the clients, and maintain a strong commitment to service.

Second, the student body should be expanded to include those other than the typical undergraduates. As the technology in agriculture continues to increase, there is a need for well-educated and skilled agriculturalists. We need to reframe our concept of the undergraduate, and undergraduates need not be limited to those between the ages of 17 and 21.

Third, the attraction of minority students to colleges of agriculture must be enhanced. Despite substantial recruiting efforts at Cornell, the proportion of minorities in the College of Agriculture and Life Sciences is about 13 percent, compared with 24 percent in the College of Arts and Sciences. There is a significant need to communicate the changing image of agriculture to minority students. We submit that, not only for minority students but for all students, if agriculture is understood to be a science-based program that focuses on the issues of biotechnology, environment, energy, information technologies, rural communities, and economics, the best students will be attracted to our colleges.

Challenges for Graduates of Colleges of Agriculture

The food and agricultural problems of today and through the 1990s require integrated multidisciplinary efforts. There is an ever-increasing need to develop comprehensive systems that integrate financial and marketing options, production technologies, and resource management practices that maintain a clean environment and that are socially responsible. During the twentieth century, agriculture was transformed from having an early focus on production, which was further enhanced by mechanization, followed by gains through chemical-based technological processes, which has led to what Hardy (1988) calls the "era of biology." This era of biology began in about 1950 and has grown to become the dominant science of the 1990s and for the twenty-first century. Graduates of colleges of agriculture must deal with the new science, which consists of biotechnology, information technologies (computing, robotics, microelectronics), concerns for the environment, energy conservation and use, new materials (both food and nonfood products), trade and policy issues, and human capital. These are great challenges indeed; and colleges of agriculture must prepare students to address the challenges of change, conflict, communication, cooperation, competitiveness, and control.

Change

It has been said that there is nothing as constant as change. As environmentally conscious farmers face the world today, they fully recognize the challenges of change. The heightened concern for food safety and groundwater quality is forcing a change from the high dependence on chemicals that was so prevalent in the 1950s. Not only have farmers faced pressures for change from the public and its perceptions about the safety of pesticides but they have also found that pests have developed a resistance to numerous pesticides rendering them ineffective. In addition, there is another challenge for change in the concept of sustainable agriculture as it attempts to use integrated agroecosystem concepts to reduce the use of chemicals.

Not only will farmers need to make changes but there must be changes on the part of consumers as well if reduced amounts of chemicals are to be used in agricultural production. Reduced pesticide usage—at least in the short run, until alternative practices are more fully developed and proven—is likely to result in more blemishes on fruits and vegetables and increased amounts of insect parts in food. Polls of a few years ago suggested that when consumers were confronted with a choice between fruits and vegetables with blemishes but that were grown without the use of pesticides and unblemished foods produced with pesticides, they would choose the blemished products over the better-appearing foods. More recent polls indicate that the choice of foods grown without pesticides has eroded and that, because of price differences, consumers are not purchasing organically grown products like they did earlier.

Conflict

We generally seek to move from conflict to convergence, and there is a general perception that conflict is bad and destructive. However, it has been said that creativity is forged on the anvil of conflict. There is much growth and understanding that can be developed from constructive conflict. Graduates must not decry the conflicts that will exist but must seek to develop processes by which groups listen to one another and work toward common objectives.

Communication

Too often, communication among industry, regulators, researchers, extensionists, and farmers has been poor or nonexistent. Unfortunately, it is much easier to talk about communication than it is to develop mechanisms by which real communication can take place between these groups, which have their own special interests. Communication of science to the general public is difficult at best and is increasingly difficult when continually mixed messages emanate from regulators, industry, and scientists. Graduates must be able to cut through the "hype" and work toward a true dialogue between concerned parties. As succinctly stated by Barker (1990), "all need to speak, all need to listen, all need to learn."

Cooperation

Cooperation and coordination must be developed among farmers, state legislatures, departments of agriculture, departments of health, departments of environmental protection, and educational institutions to develop an agenda that addresses the numerous agricultural issues. Graduates must work with all of these organizations to stress the concepts of agricultural and ecological literacy so that future generations can understand the important role of agriculture, economically and environmentally, in the United States.

Competitiveness

During the 1980s, as a result of global economic issues in combination with U.S. and foreign agricultural policies and programs, there was a period of crisis for agriculture. Increased competitiveness and enhanced profitability in the development of new technologies that improve production efficiency and the quality of products must be addressed to maintain the viability of U.S. agriculture. It is clear that globalization not only exists in agriculture today but must also be recognized as a principal driving force in the future. Graduates must understand the nature of today's global markets and the internationalization that drives the food and agricultural system.

Control

The issue of control or regulation of agricultural practices is an ever-increasing and potentially contentious issue. Who will control the agricultural practices, and at what level is it necessary for control measures to be triggered? It is not unreasonable for the public to be confused when the U.S. Environmental Protection Agency identifies a chemical as a carcinogen and states that it must be banned, but not for another several years. The concentration of chemicals is a key element here. With modern instrumentation we are able to measure ever smaller quantities of pesticides and chemicals in our water and food than we could before these modern means of measurement were available. This has led to a tendency to consider the detection of a pesticide or a chemical as being a problem when, in fact, the rational approach is to compare the concentration against the threshold that has been established for human consumption. The process is made ever more difficult by the lack of good data about the toxicity of a pesticide or chemical and the determination of how much gets into the food supply. Graduates must be prepared to address these questions.

Agriculture: An Integrated System

There are agricultural sciences, but agriculture itself is not a science. From our conceptual definition of agriculture and a consideration of extensive discussions and ideas, we suggest that agriculture is an integrated system. This system is conceptually illustrated in Figure 9-1, which embodies the following concepts:

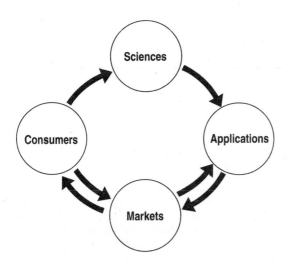

Sciences

Mathematics

Physical Sciences
  Chemistry
  Geology
  Physics

Biological Sciences
  Basic Biology
  Biochemistry
  Ecology
  Genetics
  Microbiology
  Physiology

Engineering Sciences
  Computing
  Electronics
  Fluid Mechanics
  Transport Processes
  Materials
  Thermodynamics

Social Sciences
  Communication
  Economics
  Psychology
  Sociology
  Statistics

Humanities
  Government
  History
  Linguistics

Applications

Animal Production
Animal Systems
Applied Economics
Bioprocessing
Biotechnology
Business and Management
Energy
Environment (air, land, water)
Food Processing
Forestry
Human Resources
New Products
Non-Food Products
Pest Management
Plant Production
Plant Systems
Rural Sociology
Safety
Waste Management

Markets

Agribusiness
Chemicals
Construction
Consulting
Educational Institutions
Electrical and Electronic
Energy Companies
Farm Equipment
Financial Institutions
Food Companies and Equipment
Government
Greenhouses and Nurseries
Marine
Ornamentals (turf)
Pharmaceuticals
Production Agriculture
  (farming)
Regulatory Agencies
Software
Soil Management
Space
Technology Transfer
  (extension)
Wastes
Water Management

FIGURE 9-1 Agriculture as an integrated system.

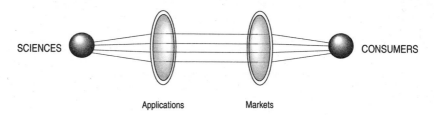

FIGURE 9-2 Thin-lens model of agriculture as an integrated system.

• Basic sciences provide the fundamental basis on which knowledge is applied.

• The application of scientific knowledge to the food and agricultural system is the focus of colleges of agriculture as it is typically played out through their respective departments.

• Markets represent areas of technology development where graduates apply their professional expertise.

• The ultimate beneficiary of the products developed in the marketplace is the consumer.

We ask the reader to view this circular model as a conceptual representation of agriculture as an integrated system and to try to refrain from focusing on the specific details of or the entries missing from the diagram. The important point is that this model displays agriculture as a system in which the basic sciences are applied for the development of the markets that serve the needs of consumers. Colleges of agriculture play that important role of converting science into applications that are usable by the market segment. The double arrows suggest the feedback of markets on applications and of consumers on markets. One might well debate whether consumers have even a weak feedback on the sciences; however, this weak feedback of consumers on the sciences provides a nice symmetry.

The thin-lens model illustrated in Figure 9-2 suggests that colleges of agriculture act like an optical device. The pair of thin lenses (applications and markets) acts to focus the source (science) on the focal point (consumers). At the same time, colleges of agriculture transfer and transform fundamental knowledge to a form that is usable by the market.

Figures 9-1 and 9-2 are presented in the spirit of this chapter, which has sought to communicate fundamental concepts and serve as a basis for discussion. In either model, colleges of agriculture are critical elements in the transformation of scientific knowledge for the benefit of society, and colleges of agriculture are key elements in the integrated system of agriculture.

References

Barker, R. 1990. Concluding remarks. Pp. 27–34 in Agricultural Biotech-
nology: Food Safety and Nutritional Quality for the Consumer. NABC
Report 2. Ithaca, N.Y.: National Agricultural Biotechnology Council,
Cornell University.

Hardy, R. W. F. 1988. Agricultural biotechnology and the environment.
Pp. 30–36 in Proceedings from the Governor's Conference on Agriculture and the Environment: A Convergence of Interests. Department of
Agriculture and Markets and Department of Environmental Conservation, Albany, N.Y.

Kupfer, A. 1988. Managing now for the 1990's. Fortune 118(7):44–46.

Muir, J. 1911. My First Summer in the Sierra. New York: Houghton Mifflin.

Educating a Culturally Diverse Professional Work Force for the Agricultural, Food, and Natural Resource System

William P. Hytche

Demographics indicate that by the year 2000, nonwhites will make up 29 percent of the new entrants into the labor force and that nonwhites, women, and immigrants will make up more than five-sixths of the net additions to the work force between 1985 and 2000 (Johnston and Packer, 1987). Irrespective of gender, by the year 2000, the minority population will be predominantly black in most states except those in the Southwest and far West, which will be primarily Hispanic. We must ensure that our work force has qualified leaders, decisionmakers, and skilled workers and scholars who can think critically, solve problems, communicate effectively, and help the United States maintain its agricultural, technological, and manufacturing superiority. This implies that the training of minorities must assume greater significance in our colleges and universities if we intend to maintain our scientific expertise. However, this cannot be done in isolation. Individual institutions may embark on special initiatives, but they are usually tentative and of short duration. This subject is not new to me, since I have been speaking about educating blacks for years. Once, this was from the standpoint of fairness and social justice. Today, it is an urgent matter of national security.

A national initiative focusing on minority human expertise development must be our priority for the agricultural sciences if this discipline intends to play its role in maintaining a stable professional work force. The urgency for our prompt and decisive action comes at a time when the climate for training minorities is not at its best. For example, the Johnson Foundation (*Wingspread: The Journal*, 1990) revealed that:

- more black males are in jail than in college;
- a black man's life expectancy is 6 years shorter than a white man's; and
- the leading causes of death among black men ages 18 to 35 years are homicide, suicide, and lung cancer.

Young black men need help if they are to escape being one of these statistics. Since it is more costly to keep an inmate in jail than it is to keep a student in college, why is so little being done to avert the crisis of the black male? Suggestions regarding the extinction of the black male are not a dramatization but rather a reality. I challenge you to look at minority student enrollment on campuses and to compare the male-to-female ratio with that of 10 years ago. There is a crisis in the U.S. higher education system as it relates to the training of minorities. There are divergent opinions regarding the causes and prevention. We have had enough blame to go around. Parents blame teachers, and teachers blame parents; some blame the government, the school and college systems, the police, and the courts. Some point to a system of social failure and moral decay.

As educators, we must lead the way to a new world order: a new world order of targeting our minority youth toward excellence. Who is better equipped to light the torch and lead the way? Educators, just as they did through the land-grant movement and later through the Industrial Revolution. Educators must be the cornerstone for this new world order. The agricultural, food, and natural resource system has been a pioneer in the past, so responding to new challenges is nothing new for the land-grant community in general.

Some may ponder why this is a new challenge, because we have been training minorities all along. It will be a challenge if we are to maintain our competitive edge. It will be a challenge to attract qualified minorities who still regard the agricultural, food, and natural resource system as the primary vehicle for their prior enslavement. Also, with the significant increase in the number of 18-year-olds who are minorities, other disciplines will become more actively involved in competing for the high achievers.

What will be the carrot in this highly competitive arena? Schools of medicine, engineering, law, dentistry, etc., will promise lucrative postcommencement careers. Athletics departments will promise good scholarships with the possibility of lucrative professional contracts. What will be the carrot for colleges of agriculture? Will it be a $1,000-per-year scholarship, a 5-year degree program, and maybe a job selling feed and fertilizer?

What can colleges of agriculture do? I will try to identify some critical points and initiatives that, from my perspective, are crucial if educators intend to take up the torch. Some of them may be ongoing, but the need is for a national initiative, which can be classified into four segments: early intervention (pre-high school), precollege intervention (high school), college, and postbaccalaureate programs. I will focus on each of these.

Develop and Implement Early Intervention (Pre-High School) Programs

The agricultural, food, and natural resource system must begin recruiting students long before high school. If not, there will be no high achievers left. Recruitment must be through innovative program initiatives in grades three through eight.

Special Skills Sessions

Nutrition, agronomy, animal science, natural resources, forestry, etc., can be fused into special science, math, and reading skills sessions for third to eighth graders. One could also offer an orientation to agriculture through computers and laboratory instrumentation. We need to educate the elementary textbook writers so that they know something about the agricultural sciences.

Saturday Academy

A faculty member could devote 3 or 4 hours one Saturday per month to bring at-risk minority students onto a campus and expose them to some of the activities of the Agriculture in the Classroom program of the U.S. Department of Agriculture. Through a faculty rotation system, no one would be occupied more than twice in a year, unless they have a specific desire to do so. This time could also be used for students to conduct independent science experiments, stimulate their thinking, and enhance their interest in and perception of agriculture.

Motivational Sessions

Faculty could conduct motivational sessions with minority students. The lives of many of our minority youth are devoid of positive experiences. Sessions in goal setting, leadership development, and social values could prepare them for outstanding future careers. Many black youth need constant reminders that there are opportunities for them.

High School Intervention Programs

Adopt a High School

Many minorities still perceive agriculture as farming. Through an infusion of agriculture, food, and natural resources into the high school curriculum, some of this stereotyping can be avoided. Faculty may choose to be guest lecturers at a high school twice a semester or may substitute for a science teacher once a semester.

Minority Research Apprenticeship Program

In this program, minority high school juniors and seniors who are in the upper third of their classes are invited to spend the summer on campus with bench scientists. They do independent science projects and computerized literature searches and are provided a laboratory science orientation to the agricultural sciences. They are also paid a stipend.

Summer Scholars Program

Outstanding students are invited to spend time on campus for 1 to 2 months to participate in some agricultural science activities for college credit. This is an opportunity for faculty to observe student performance; for students to establish contacts, develop mentor relationships, and decide on career options; and for universities to award scholarships.

Preparatory Summer Internships

Most of us design summer programs to attract the high achievers. These students are usually in the top 30 percent of their classes. What do we do with the other 70 percent? A major flaw in our precollege program initiatives is our intense competition to attract only the upper echelon. We must implement programs, for example, summer work experiences, to expose these students to the food and agricultural sciences. Many of our youth are late bloomers who require a little push or some incentive to excel. Minority students should never be categorized into those who can make it and those who cannot. If they are, educators should challenge themselves to work with the students who they think cannot make it. Many of these students have never been told by anyone that they have potential or that they can do it. We can adopt future scientists from this group and cultivate the philosophy that high achievers may come from impoverished single-parent homes. Paths to success may differ; what matters is solid preparation coupled with determination and a positive self-image, all to which educators can contribute.

Reevaluation of Entrance Requirements

Many minority students are casualties of standardized testing, inner city myopics, and the perils of growing up male and black. The scores that students achieve on the Scholastic Aptitude and American College tests often do not reflect the academic potential of students, particularly black students. The summer scholars program mentioned above could provide the opportunity for a more effective evaluation of that lower 70 percent of the students.

High-Profile Recruitment and Marketing Initiatives

Most educators are engaged in some initiatives to recruit minority students. We need, however, to go beyond these independent activities and embark upon a national advertising program similar to that done by the National Science Foundation, the U.S. Army, and others. We should target minority audiences with a specially designed, high-profile, nationally televised advertisement. We should develop appropriate career-oriented recruitment brochures and videos that could be distributed in the high schools. We should develop a network of black churches to gain access to parents and students. We should develop cartoon strips, newspaper advertisements, and other resource materials. We should try to access nontraditional media such as Black Entertainment Television on cable television and *Jet*, *Ebony*, and *Essence* magazines. Above all, we should not rely exclusively on recruitment contact and referral slips, but rather, we should establish and maintain constant contact with prospective students and parents through letters, postcards, telephone calls, and when feasible, personal contact. The philosophy of the latter is that, just as coaches pay personal visits to the parents of athletes, would it not be possible for educators to do the same for a prospective student with a 4.0 grade point average? He or she is an academic superstar.

Reevaluation of Vocational Education Programs and Creation of More Agricultural High Schools

Colleges and universities need to assist in the revision of curricula in vocational education programs. The administrators and teachers of these programs are our graduates, so do not say this task cannot be done. Programs in the agricultural, food, and natural resource sciences at the secondary school level should be science based and not purely vocational.

College Programs

When one examines the record, one is puzzled by our achievement, or lack thereof, in retaining minorities until they graduate. Over the past 30 years, the 1862 land-grant universities have attracted 65 percent of the minorities enrolled in the agricultural, food, and natural resource sciences but have graduated 35 percent of the minorities receiving undergraduate degrees in these areas. I am not sure how this compares with other disciplines, but it is certainly a reflection on the admissions process, the curricula, or the advising process.

Relevant Curricula

First and foremost in our college program initiatives to educate a culturally diverse professional work force for the agricultural, food, and natural resource system is a broadened, more inclusive relevant curricula. According to the December 5, 1990, *Chronicle of Higher Education*,

> Since 1978 undergraduate enrollment in the agricultural sciences has decreased approximately forty-five percent. Courses in natural resources, general agriculture and plant and soil science have lost more than half their students. Although the number of undergraduates enrolling in agriculture in the last two years has rebounded, educators should not be lulled into a false sense of security. Our students have shifted to programs such as biochemistry, agricultural economics, agribusiness management, nutrition and other basic sciences (Gwynn and Thompson, 1990:B2).

These trends demand a revitalization of the curricula and fundamental changes in the preparation and career orientation of students.

Effective Mentoring Program

We have many first-generation minority college students and will continue to have them for some time. Many lack appropriate role models, and their history and knowledge of agriculture are reminders that they were brought here in chains as forced immigrants and were required to work in the fields. Many are without supportive families, and many who have graduated indicate that their greatest fortune was finding a mentor with whom they built a positive relationship. The fragility of the family and the community and the prevalence of crime, drugs, and alcohol dramatize the need for positive role models in colleges. Role models and mentors can be the difference between success and failure. As we attempt to develop mentoring programs, we must realize that not all of us can be mentors. Some of us have low tolerance levels, while others have little respect for minorities. Mentoring cannot be thrust upon everyone. It must be in the heart. If there exists an intolerance for minorities, be open and do not accept the responsibility of mentoring. Deans and other administrators must, however, be aware that this intolerance can be pervasive throughout one's entire job responsibilities. Classroom performance should, therefore, be monitored. Do not treat lightly complaints from minority students regarding racism in the classroom. Complacency or the failure to respond could be the decisive element in charting the future of a minority student. As teachers and mentors, we possess the power to make lives joyous or miserable. We can be the tool of torture, or we can be the instrument of inspiration; therefore, do not humiliate, humor, or hurt those with whose future lives you have been entrusted.

Financial Assistance

The escalating cost of higher education prohibits many minority students from considering higher education. A recent pronouncement by the U.S. Department of Education regarding minority scholarship programs will make it more difficult for many minority students to afford college. Just as athletic programs can develop attractive financial packages to attract students, so can the agricultural, food, and natural resource system. We must pursue alumni, the agribusiness sector, corporations, faculty, and federal agencies. Just as colleges and universities have been innovative in coercing the Kellogg Foundation, Du Pont Co., Monsanto Co., Pepsi Cola, Coca Cola, the Carnegie Foundation, R. J. Reynolds, and others to give multi-million-dollar facilities and equipment, so can they be innovative in soliciting scholarship funds to maintain and enhance our human potential in the agricultural, food, and natural resource system. Innovative cooperative education and paid internship programs can also contribute to the financial package of the student and play a significant role in attracting and retaining minority students. Cooperative education and internships can also play a significant role in the student's career decision-making process.

Liaison Relationships

We should establish liaison relationships between the 1862 land-grant institutions and institutions with significant undergraduate minority enrollments. I refer particularly to the 1890 land-grant institutions. This can be accomplished through innovative summer internship programs and/or joint research activities in which minority students could participate. This is discussed in the next section.

Postbaccalaureate Programs

Although we have recovered some of the undergraduate enrollment that was lost in the early 1980s, the minority student enrollment in graduate programs is still on the decline. The lack of mentors at schools that offer most of the graduate programs may be a contributing factor. Few new minority doctorates are being produced, so existing faculty move around.

The 1890–1862 land-grant institution liaison relationship should be strengthened, as mentioned above, and we should help to develop and encourage more master's-level programs at 1890 land-grant institutions.

It is our belief that many minority students will attend only predominantly minority institutions. If minority institutions did not exist, many minority students would not attend college. This is true even for majority students. This is a basic fact of life. The integration of minorities in predominantly white institutions or the elimination of predominantly black institutions will not change this. It has been said that the issue of access is not only who goes to college, in terms of numbers, but who goes to which college. The thought of attending a larger, predominantly white institution creates an almost insurmountable hurdle for blacks. Most of the historically black colleges and universities are one-third the size of the average majority-race institution; therefore, the climate of a small community, family spirit, and individual attention exists at minority institutions, and minority students feel at home. The creation of more master's-level programs at these institutions would further ease the transition to the larger institutions and would create more graduate students.

Conclusion

In this chapter I have discussed only the education of blacks and other minorities, when the topic was to be educating a culturally diverse professional work force. My assumption is that we are doing a good job of educating the majority population. We have also seen tremendous strides by women and Asians, but blacks continue to lag behind in terms of the percentage of college graduates in the sciences, including agriculture, food, and natural resources. I can sum up this chapter by paraphrasing the leadership actions developed by the National Task Force for Minority Achievement in Higher Education established by the Education Commission of the States for achieving campus diversity. Colleges and universities must reduce barriers. This is accomplished by recruitment, financial aid, and admission policies. In addition, although I did not emphasize it here, the time of day that courses are offered is also important to reducing barriers.

Colleges and universities must provide students with help. This is accomplished through outreach, mentoring, advising, and the climate on campus. Colleges and universities must improve learning. This can be accomplished by student assessment, learning assistance, the curriculum, and teaching practices.

The bottom line is simply that we must be innovative in our approach, because we cannot do the things we have been doing—they have not worked.

References

Gwynn, D., and E. O. Thompson. 1990. Agriculture schools must broaden their curricula to attract new groups of students. Chronicle of Higher Education, December 5, 1990, pp. B2–B3.

Johnston, W. B. and A. H. Packer. 1987. Workforce 2000: Work and Workers for the 21st Century. Indianapolis, Ind.: Hudson Institute.

Wingspread: The Journal. 1990. Young black males need help. Wingspread: The Journal 12(2):1.

Scientific Literacy: The Enemy Is Us

Robert M. Hazen

Pick up a newspaper any day of the week and you will find a dozen articles that relate to science and technology. There are always stories on the weather, energy, the environment, and medical advances—the list goes on and on. Is the average American prepared to understand the scientific component of these issues? I am afraid the answer, in almost every case, is no.

In this chapter, which is adapted in part from a previous report (Hazen and Trefil, 1991c), I will first describe my perceptions of the nature and origins of the national scientific literacy problem and then propose a solution that U.S. science educators can implement with reasonable ease and with a sense of optimism.

I want to share a few horror stories about the deficiencies in U.S. science education. Many others have presented similar examples (see, for example, American Association for the Advancement of Science [1989], Bishop [1989], and National Research Council [1990]). The series of personal anecdotes I describe may, at first, seem very different and perhaps unrelated. Yet, I think these separate incidents can be tied together to create a much larger and more sinister picture of the way science is taught in the United States.

The first is the story of my daughter Elizabeth. She is a seventh grader and about as bright as any student you would ever want to meet. Year after year she has been required to memorize lists of vocabulary words in science. Her sixth-grade science teacher introduced her to terms like *batholith*, *saprophyte*, *nekton*, and *hyphae*. She had to define Hertzsprung-Russell diagrams and Fahnestock clips, all in the sixth grade. Ten- and 11-year-old children are being taught vocabulary that the average doctoral-level scientist does not know.

Elizabeth also had to memorize the names and accomplishments of 40 African-American scientists. That is a laudable thing to teach, but rather than learn about these individuals as real people who had aspirations, who had tremendous roadblocks to their careers, and who ultimately triumphed, she had to memorize a list of 40 names on one side of a piece of paper and 40 accomplishments on the other: the students had to match the pairs. For example, she had to remember that Caldwell McCoy worked to create energy from magnetic fusion. Unfortunately, the children in the sixth grade were never told about nuclear fusion, nor had they the slightest notion of what magnets had to do with fusion energy.

My daughter is very good at memorizing things, and she did quite well in the class. And now she hates science.

My son Benjamin is in high school, and he is also a fine student. Like most high-school students, he takes science, with weekly laboratories where everyone mixes the exact same amount of the exact same chemicals and gets, one hopes, the exact same results. The standard titration experiment (the one where solutions turn pink) was one of his required exercises. If Ben wanted a good grade, he had to get the right answer; there was no margin for error, no chance to think about other experiments, and no opportunity to see what would happen if he mixed things in slightly different ratios.

Ben is pretty clever and he figured out how to do the experiments. When I asked Ben what an acid or a base is, he looked at me with a blank stare—he did not have the vaguest idea. Now, Ben hates science, too, and he asked me recently, "Why would anyone want to do this sort of thing for a living?" That is a tough question for a scientist to be asked by his child.

I now move to the college level. The scene is Harvard University on the festive 1987 graduation day as recorded in the film *A Private Universe* (Pyramid Film Video, Santa Monica, Calif.). A group of seniors, about 25 in all, was asked, "Why is it warmer in the summer than it is in the winter?" Granted, it was graduation day, everybody was in their festive robes, champagne corks were popping, and so forth. Of those 25 graduating Harvard seniors, however, only 2 answered correctly. The vast majority said it is warmer in the summer because the earth is closer to the sun, an explanation that should make any elementary-school science teacher cringe.

A similar survey at my own university, George Mason, revealed that more than half of the graduating seniors could not describe the difference between an atom and a molecule. There is no doubt that we are producing a generation of college graduates who do not know the most basic facts about science.

What about working scientists, those of us who are in the elite? I did an informal survey among two dozen of my colleagues in the earth sciences. Each was asked a very basic question in biology: "In simple qualitative terms, what is the difference in function between DNA and RNA?" (They are two fundamental molecules present in all life. The first carries the genetic code; the second interprets it.) Of two dozen geologists, only two could answer the question, and both of them were biogeochemists engaged in a study of fossilized DNA remnants in rocks.

To even out the record, I asked a group of biology professors, "What is the difference between a semiconductor and a superconductor?" Apart from a bad joke about the local symphony orchestra, not one biologist could respond to that question.

One might think that my informal survey included only second-rank scientists, that the cream of the crop really possesses a broad general scientific knowledge. Unfortunately, it is not true. I recently engaged a Nobel Prize-winning chemist in a conversation about the 1989 San Francisco earthquake. In the course of that discussion I mentioned plate tectonics. He looked at me and said, "What's plate tectonics?" When I told him of the transformation that has taken place in the earth sciences in the last few decades, he seemed only politely interested. He certainly was not the slightest bit concerned that one of the most important discoveries in our understanding of the planet had completely passed him by.

Each of these stories about science is disturbing in its own right. But taken together, they present a truly bleak picture of the state of American science education. At every level we are failing to provide students with the information they need as citizens. Elementary school children are learning that science is difficult, boring, and irrelevant to their day-to-day lives. College students are graduating without knowing the most basic concepts about their physical world. And even working scientists are often scientifically illiterate outside their own narrow specialty.

What is going on? Why has the system failed so many people? My colleague Jim Trefil and I believe the answer boils down to the misdirected priorities of scientists ourselves. If you listen carefully to most scientists when they talk about the scientific literacy crisis, what they really mean is that fewer and fewer people are becoming scientists. At present, about 1 percent of the U.S. population are scientists, and as far as most scientists are concerned, the other 99 percent are of little concern.

In a sense, it seems that most scientists' educational objectives are to try to make everyone a scientist. We start out with that as a goal, and then we weed out the unworthy. We subject children at earlier and earlier ages to more and more big words, mathematical rigor, and experimental abstractions. Then, as children get turned off, a significant number of educators advocate teaching even more big words at an even earlier age to reverse the trend. They press for more mathematics and additional fancy experiments. Is this strategy going to attract more children to science? I think not.

I attended a junior high school physics class recently where this

97

problem was highlighted. The teacher faced the classroom and said, "Today we are going to talk about gravity." Then he turned to the blackboard and, as he wrote, said, "This is the equation for gravity," and proceeded to analyze the variables.

What a terrible thing to do to a group of 12-year-old children. Gravity is jumping off a chair, dropping a ball, or Michael Jordan doing a 360-degree slam dunk. Look at your day-to-day lives and see how often gravity comes into play. Children need the chance to recognize that there are only a few physical forces that control our lives. Then, if they are interested enough—if they want to know how they might get a satellite into orbit—you can write down equations and apply them. Until children understand, however, that gravity is a force that affects them every day of their lives, the equation means nothing. It is a needless abstraction.

How many students in that class will decide to devote their lives to science? The chances are, none. Our biggest mistake as educators is that from the earliest grade we try to make all of our students miniature scientists. If they succeed at one stage, then they go on to bigger and harder science. Eventually, at some level, almost all drop out—almost all have failed. Scientists have become like an elite priesthood of knowledge: Only the worthy who have completed the rites are allowed in. No wonder so few Americans want to become scientists, and no wonder there is so much mistrust of science and scientists. What other academic field makes the majority of students feel like failures?

The situation is, if anything, worse in U.S. colleges and universities than it is in the elementary schools. Two problems pervade the organization and the presentation of science at the college level (Hazen and Trefil, 1991a,b). First, almost all science courses, even those for first-year nonscience students, are geared for science majors. Such courses are intended to give a foundation for further study. As a result, they rarely provide any sort of overview of science and do little to foster scientific literacy among scientists or nonscientists. Second, science courses rarely integrate physics, chemistry, biology, and geology. Students must take courses in at least four different science departments to gain even a basic level of literacy in all the sciences.

A number of very successful departmental courses have addressed the first problem. Offerings like Physics for Poets—courses designed specifically for nonscientists—do a very good job of introducing a specific discipline. But these courses still leave the nonscience major with exposure to only one branch of science. There is little chance that a student in Physics for Poets will learn about modern concepts of genetics or plate tectonics. In short, the science curricula of most colleges and universities fail to provide the basic science education necessary to understand the science and technology issues facing this society.

The ever-increasing specialization of science has as one of its sorriest consequences the fact that most working scientists are themselves scientifically illiterate. I am a perfect example. The last time I took a course in biology was in ninth grade, long before the modern developments in genetics appeared in any textbook. In college I studied lots of earth science, and in graduate school I studied lots more. But no one ever suggested that I take a biology course, and I was not likely to waste my time reading about something that my professors did not expect me to learn.

From that distant day in 1962 when I last dissected a frog until just a couple of years ago when I began teaching general science, I was about as ignorant of modern biology as it was possible to be. We are a society that has lost the ability to teach general science at any level, because there is almost nobody who knows enough to teach it.

The United States is number one at producing specialized professional scientists, scientists that have the research skills, the insights, and the abilities to do top-notch research to lead our country in technology and to lead the world in teaching new doctorates (even if most of those students come from other countries). Our specialization has come at a price. National science leaders, the people who are best at playing the research game, have fostered an education policy so concerned with producing the next generation of specialized scientists that the education of the 99 percent who are not going to be scientists has gone by the wayside. And this policy has backfired, because it is turning off U.S. science students in record numbers.

The problem is daunting, but there is a realistic and straightforward solution. Our specific solution addresses the scientific literacy crisis at the college level, but the same principles could be extended to both pre- and postcollege learning.

Jim Trefil and I have developed a course for undergraduate non-science majors at George Mason University. The university's science core curriculum committee has recommended that all non-science majors take this course in their first year; that course should be followed by a two-semester laboratory course in physics, chemistry, geology, or biology. In this way, every student receives an introductory overview that is followed by a specific science course that gives them some experimental rigor, that introduces laboratory technique, and that examines the analytical process in doing specific scientific experiments. This pairing starts with an integrated view of physics, chemistry, biology, and earth science and then explores one subject in greater depth.

It must be emphasized that scientific literacy does not mean producing more scientists. That cannot be the goal of effective general science education. The principal objective is to give all Americans the information they need to understand the kinds of technological and social issues that confront us every day. This knowledge is not the specialized stuff of the experts; it is an eclectic mixture of vocabulary, general principles, some history, and some philosophy. This core knowledge changes only gradually with time, in contrast to the constantly shifting events in the news. If students can take a newspaper article about genetic engineering, the ozone hole, or chemical waste and put those in a meaningful context—if they can treat science in the same way that they treat any of the other pieces of information, like sports, business, or economics, that comes their way—then they are scientifically literate.

The major challenge we face as science educators is that most societal issues concerning science and technology require a very broad range of knowledge. To give just one example, consider nuclear waste. To understand nuclear waste, you need to know how nuclei decay to produce radiation (physics), how radioactive atoms interact with the environment (chemistry), how radioactive waste can enter a geologic system (earth science), and how radiation affects living things (biology). Students cannot begin to understand the nuclear waste problem if they have studied only physics, chemistry, or biology. Similarly, many other issues, including global warming, space research, alternative energy resources, and medical technologies, depend on a whole spectrum of scientific concepts. Scientifically literate scientists and nonscientists alike need to understand a little bit of several disciplines to cope with these issues.

Science forms a web of interconnected knowledge. The key to producing scientifically literate graduates is to recognize that there are a few basic overarching ideas in science, ideas that connect all of our physical experience. We live in a universe of matter and energy—that is all there is with which to play the game of life—and science is simply the set of rules about how matter and energy behave. Once students are introduced to these basic principles about matter and energy, they begin to see science as a special way of understanding their universe. They learn to place science in a much broader human perspective.

We organize our course, Great Ideas in Science, around a series of about 20 overarching principles. The list of great ideas is neither obvious nor immutable (Culotta, 1991). Any scientist could come up with a compilation of 20 or so key concepts. Compare a dozen different lists and 8 or 10 ideas will appear just about every time. Newton's laws of motion, the laws of thermodynamics, and the concept of the atom are common to all scientific disciplines, for example.

The most basic principle, the starting point for all of science, is the idea that the universe can be studied by observation and experimentation. It is remarkable how many students—even science majors—have no clear idea how this central concept sets science apart from religion, philosophy, and the arts as a way of understanding our place in the cosmos.

Once students understand what science is, then they can appreciate the basic principles shared by all sciences, things that are traditionally covered in early physics courses: Newton's laws that govern force and motion, the laws of thermodynamics that govern energy and entropy, the equivalence of electricity and magnetism, and the atomic structure of all matter. These are not abstract concepts. They apply to everyday life, explaining, for example, the compelling reasons for wearing seat belts, the physics of a pot of soup, and the contrast between static cling and a refrigerator magnet. In one form or another, all of these ideas appear in virtually every elementary science textbook, but often in abstract form. We should strive to make them part of every student's day-to-day experience.

Once the general principles have been laid down, we can look at specific natural systems such as galaxies, the stars, the earth, or living things. For each of those systems, additional principles must be stated. In astronomy, for example, students learn that stars and planets form and move according to Newton's laws, that stars eventually burn up according to the laws of thermodynamics, that nuclear reactions fuel stars by the conversion of mass into energy, and that stars produce light as a consequence of electromagnetism.

Two basic ideas—plate tectonics and earth cycles (rock, water, and atmosphere)—unify the earth sciences. The laws of thermodynamics decree that no feature on the earth's surface is permanent. This principle can be used to explain geologic time, gradualism, and the causes of earthquakes and volcanoes. The fact that matter is composed of atoms tells us that individual atoms in the earth system—for example, in a grain of sand, a gold ring, or a student's last breath—have been recycling for billions of years.

Living things are arguably the most complex systems that scientists attempt to understand. We identify five basic principles that apply to all living systems: all life is based on chemistry, all life is made up of cells, all life uses the same genetic code, all life evolved by natural selection, and all forms of life are interconnected as parts of ecosystems.

The great ideas approach has a tremendous advantage for students and for teachers. Any issue of scientific or technological importance can be introduced as a way of illustrating the general principles. We frequently use examples from recent newspapers: New materials emphasize how atoms combine to cause distinctive properties; environmental concerns illustrate ideas of ecology; space probes raise questions about the planets. It is likely that the issues that loom large today—AIDS, drug abuse, and the hole in the ozone layer—may seem insignificant in a few years, while new issues will undoubtedly come to take their place. By focusing on general principles, whatever issue comes along, the teacher can immediately integrate that into the basic framework of the course. Furthermore, each teacher can choose examples to suit his or her interests and style, and the underlying principles will remain the same.

Another important benefit of the great ideas format—one that has special relevance to agricultural education—is that many important technological fields are poorly represented by traditional science departments. Computer and information technology, brain research, and medical science often are not integrated into traditional courses in chemistry and physics. Agriculture and natural resources represent other fields in which the traditional physics, chemistry, or biology courses do not touch on important issues. By teaching a general principles course in science, every illustrative example can be chosen from a favorite discipline, without sacrificing generality.

We also are able to look at science from a rich variety of viewpoints using these general principles. For example, Newton's laws of motion are among the central ideas in science, but students should also learn things like when Isaac Newton lived, how he incorporated the earlier work of Galileo and Kepler into his system, and how Newton's work influenced the philosophy of the Enlightenment. Newton's laws can be used to illustrate such practical examples as why people should wear seat belts, the launch of a space shuttle, or even the difference between football linemen and quarterbacks. Newton also provides an excellent starting point to discuss the relationship between science and technology, the importance of experimentation in science, and the scientific method—all key concepts that are not covered in most science courses.

The most frequent objection to the Great Ideas in Science course is "No one will be able to teach it." Such a criticism of our course is, in itself, a serious indictment of the science education system. If professional science educators are unable or unwilling to learn the most basic principles of other scientific fields, then how can we expect nonscientists to gain any level of scientific literacy themselves? If a physics teacher refuses to learn biochemistry or biologists shun plate tectonics, why should students care at all about these subjects?

Ideally, one faculty member should teach the entire course. We have found that many faculty members at George Mason University are eager to do this. During the 1990 spring semester, eight faculty members representing all the science departments attended the course and are now ready to teach it themselves. None of us is an expert in all the fields covered, and student's questions often leave us stumped. But it provides a valuable lesson to the class when you say, "I don't know the answer to that question, but I will find out." What better way to emphasize to students that science is an ongoing process of learning?

An obvious alternative is to have several faculty members teach in their own disciplines; thus, chemists, geologists, and biologists could stay close to their own turf. We discourage such an approach for several reasons. One key theme of the course—that the sciences are integrated and they form a seamless web—is lost. Specialists tend to slip into confusing jargon and dwell on unnecessary detail, thus defeating the purpose of the overview. Finally, students may well ask why they must master a range of scientific topics when the faculty members appear to be unwilling to do so.

Student response to this course has been overwhelmingly positive. Students complete detailed course evaluations at the end of each semester, and they have always placed the course in the top 10 percent of course offerings at George Mason University. Many students who are nonscience majors have remarked that the course is the first one to make science seem relevant to their everyday lives. Surprisingly, many science majors said it was the first time they understood what they were doing as scientists; as specialized science majors, they had never seen the big picture.

Science educators throughout the country have created a system that alienates science students from their earliest years. At every grade level the accumulated vocabulary and the mathematical abstraction winnows out students. By returning to general science courses for all students, colleges can, in some measure, reverse the trend. Our goal must be to produce college graduates who can see that scientific understanding is one of the crowning achievements of the human mind, that the physical universe is a place of magnificent order, and that science provides the most powerful means for discovering knowledge that can help us to understand and shape our world.

References

American Association for the Advancement of Science. 1989. Project 2061: Science for All Americans. Washington, D.C.: American Association for the Advancement of Science.

Bishop, J. 1989. Scientific Illiteracy: Causes, Costs and Cures. Washington, D.C.: American Association for the Advancement of Science.

Culotta, E. 1991. Science's 20 greatest hits take their lumps. Science 251:1308–1309.

Hazen, R. M., and J. Trefil. 1991a. Achieving geological literacy. Journal of Geological Education 39:28–30.

Hazen, R. M., and J. Trefil. 1991b. General science courses are the key to scientific literacy. Chronicle of Higher Education, April 10, 1991, p. A44.

Hazen, R. M., and J. S. Trefil. 1991c. Science Matters: Achieving Scientific Literacy. New York: Doubleday.

National Research Council. 1990. Fulfilling the Promise: Biology Education in the Nation's Schools. Washington, D.C.: National Academy Press.

The Priority:
Undergraduate Professional Education

Joseph E. Kunsman, Jr.

How do we educate students so that they can meet the demands of the new world? This was asked by many who attended the conference. The information presented at the conference and in this volume shows that the methods and materials are already there. It is like the man of the East who came to the monk in the marketplace and asked the way to the city. Those within hearing distance laughed. He was already there. So with us.

A recent reading gives me pause and raises a strong concern. The reading was from a letter dated 1855 and was sent by Chief Seattle to the President of the United States regarding yet another sale of Indian land to the expanding American republic. That famous letter begins with the question and comment: "But how can you buy or sell the sky, the land? The idea is strange to us" (Campbell, 1990:28).

Chief Seattle asks another question that we have avoided dealing with for nearly all of our history: "Your destiny is a mystery to us. What will happen when the buffalo are all slaughtered? The wild horses tamed? What will happen when the secret corners of the forest are heavy with the scent of many men and the view of the ripe hills is blotted by talking wires?" (Campbell, 1990:29).

Questions regarding what to do about significant change have never been a favorite of Americans, and even less so for academics.

If you are old enough, you can remember with humor and some degree of pride the early space race—the race to the moon. One morning, Americans awoke to find that a people with strange names and a foreign ideology had put a beeping piece of technology over our heads. The beeps of *Sputnik* were greeted with recriminations all around the United States. Evidently, everyone was to blame. It very suddenly dawned on us that we did not have enough mathematicians, physicists, and the like, and nobody seemed ready or able to address the need. What followed was an exciting pyrotechnic display of U.S. Army, Navy, and Air Force rockets exploding on their launchpads.

Undaunted, however, and with typical American arrogance, our president declared a space race and said we would be first on the moon within 10 years. Of course, we were first on the moon, but in 8 years.

Unfortunately, it appears that we are no longer confronted with that kind of world. There are no more decade-long races. Instead, a major ideology in Eastern Europe falls not in years but in weeks. A major war fought between the world's second and fifth largest armies takes not years but days. And when the Japanese stock market fluctuates, Wall Street and Nebraska grain farmers alike respond not in hours but in seconds.

The American distaste for planning, thinking, and problem solving will not serve us well in this antebellum world.

Joseph Campbell, who must have had Americans in mind, once said, "What men of deeds have ever listened to sages? For these, to think is to act, and one thought is enough" (Campbell, 1970:171). That must be the American way.

Business as usual is an American credo, fortified by the academic community in its resolve to staunchly conserve our heritage.

In the College of Agriculture at the University of Wyoming, a faculty task force has been reviewing our version of the undergraduate experience. We began with Paul Kennedy's epic volume *The Rise and Fall of the Great Powers* (Kennedy, 1987). The last section of that intriguing volume is titled "The United States: The Problem of Number One in Relative Decline." That section's major thesis is as follows: "(T)he only answer to the question increasingly debated by the public of whether the United States can preserve its existing position is 'no'—for it simply has not been given to any one society to remain *permanently* ahead of all the others" (p. 533). Paul Kennedy is reminding us that there is no such thing as a permanent manifest destiny, no matter how hard we wish to believe that there is.

Shaw's deadly quip fits well here: "Rome fell, Babylon fell, Scarsdale's turn will come" (Kennedy, 1987:533). But that is the point for us to ponder. Paul Kennedy explains that "the decline referred to is relative not absolute, and is therefore perfectly natural; and that the only serious threat . . . can come from a failure to adjust sensibly to the newer world order" (Kennedy, 1987:534). Who is going to plan sensibly? Obviously, this nation is counting on the products of our colleges and universities—our graduates.

Frank Newman speaks even more precisely to the dilemma in a Carnegie Foundation Special Report: *Higher Education and the American Resurgence* (Newman, 1985). Newman notes: "The economic times have changed. Ours is a more technological, more international, but most of all more dynamic world. This country's ability to compete and to lead is dependent on the nature and quality of higher education. An understanding of technology is important to graduates, but so is the capacity to take initiative, to be creative, to understand the international nature of the world, and to comprehend the need to both compete and cooperate" (p. 28).

Our educational system is the envy of all the world. Its incredible achievements need not be chronicled here. Yet, there is work to be done and important adjustments to be made. For instance, take two specific dilemmas. Many years ago we bought enthusiastically the concept of the German research university, and the research-teaching amalgam we have forged is our pride and heritage. If we did not sincerely believe that, we would not hold an undergraduate curriculum conference at the National Academy of Sciences. However, we must acknowledge that we are faced with a difficult and troubling dialectic.

I am reminded of the Indian mystic who, upon entering a temple, looked at the holy image and debated within himself whether God had form or was formless. He passed his staff from left to right. The staff touched nothing. He concluded God was formless. Then he passed the staff from right to left. It struck the image. He understood that God had form.

Of course, dialectics work well in Indian mysticism, but less so in tightly budgeted, self-indulgent, immoderate universities.

The teaching and research amalgam is our strength, but it is also our dilemma—a dilemma worthy of our immediate attention. Redressing the balance can benefit both research and teaching. Appropriate funding for both is requisite for progress. The system—and I myself—have supported the recent national research initiative for agriculture, food, and the environment. It was appropriate. An undergraduate teaching initiative is also appropriate, and it begs everyone's support.

A second dilemma seeking a remedy is that of teaching prestige and reward. Many chapters in this volume address this issue. For the noblest of professions, a noble reward is due. Ensuring an appropriate reward for our finest teachers is a difficult but achievable task. It should be pursued with vigor at both the local and national levels. Yet, another complementary note must be sounded here. Several years ago I attended a presidential conference on excellence in education. One of the highlights for me was the keynote address by Senator Richard Lugar (R.-Ind.) (Lugar, 1983). I remember that he made the point that we could no longer get by in education on the backs of dedicated missionaries. It was a popular theme at the conference, and I shared it with many of my colleagues. But over the years, the more I thought about it the less I believed it. Dedicated missionaries make the learning experience.

Someone once said that the most important ingredient in good education is the appetite—the appetite of the teacher as well as that of the student. That appetite is what got all of us to the college and university in the first place. I never cease to be amazed at those surveys that show that college and university faculty, given the choice, generally have a preference for teaching over research. It is a noble task, and we like being noble. Encouraging that nobility may be one of the easiest reforms to implement improvement in our present educational system. But we must be realistic. It is a local issue, but it needs national reform. A state university, for example, cannot do it alone. It calls for a national attitude of reallocation of resources.

Karl Brandt began the conference (see Chapter 1) with an explanation of what he believes and why he worries. As we concluded the conference, we could safely say that we believe the methods and materials for improving our undergraduate educational effort are available and workable. Yet, equally apparent is our apprehension regarding our ability to apply the proper effort and achieve the desired results. For we take note that many of these curriculum reforms are not new, just unused. These ideas have not been tried and found wanting. They have been found to be too hard and have been left untried.

In 1990, many of us were thrilled by the production on the Public Broadcast Service of Richard Wagner's Ring Cycle, *Der Ring des Nibelungen*; four operas and 16 hours of the world projected in myth and music. Woton, the one-eyed sky god, and his accompanying pantheon represent humankind. But I saw a special kind of humankind, the academic, the college professor.

The first opera in the cycle, *Das Rheingold*, is most instructive for us and serves as a fitting conclusion to this chapter. Woton, sky god and flawed hero, spends most of *Das Rheingold* attempting to manage human avarice, using his half-blind intuition. As expected, his efforts result in intrigue, hatred, and general human disorder. Frustrated by his limitations and the enormity and complexity of the task, Woton decides to abandon the real world and retreat to Valhalla, his mystic castle in the sky. *Das Rheingold* concludes with Woton and his entourage of lesser gods majestically crossing over reality on a beautiful rainbow bridge to the castle in the sky.

Those of us in academe have a magnificent history of retreating from the world and its messy problems to our mythical ivory towers. Yet, this republic of ours and its citizens have placed great, if not also slightly misplaced, confidence in us. They also invest a good portion of their fortune to support our efforts. This significant confidence and financial investment and the enormous importance of our task must elicit from us not a retreat but a herculean effort to provide the best and most applicable undergraduate education.

107

Wagner's ring cycle began with Woton's retreat from the world he helped to create, and as a result, it ends 16 hours later with *Götterdämmerung*, the twilight of the gods. This is not a comforting analogy in this most robust of times. Maybe it is best, then, that we accept our challenge: to produce educated individuals prepared, as Paul Kennedy says, for sensible planning.

This is no easy task, but we have asked the monks, "Where is the city?" The conference participants confirmed that we are already there.

Like my colleague, Karl Brandt, I believe and I worry. But mostly, I believe. Many times during the conference we heard speakers share the conundrum that history is too important to leave to historians, or that education is too important to leave to colleges of education. Using this rubric one last time, let me point out that undergraduate education in agriculture is too important to be left to deans and directors of resident instruction.

We simply cannot educate another generation in the deficient manner in which we were educated. Everyone in the agricultural community—faculty, deans, students, chief executive officers, and presidents—must become actively involved. We can do it together, but let us not leave it undone. A great republic is counting on us.

References

Campbell, J., ed. 1970. Myths, Dreams and Religion. Dallas: Springs Publications.

Campbell, J. 1990. Transformations of Myth Through Time. New York: Harper & Row.

Kennedy, P. 1987. The Rise and Fall of the Great Powers. New York: Random House.

Lugar, R. G. 1983. Keynote Address. National Forum on Excellence in Education, Indianapolis, Ind., December 8, 1983.

Newman, F. 1985. Higher Education and the American Resurgence. A Carnegie Foundation Special Report. Lawrenceville, N.J.: Princeton University Press.

13

Positioning Undergraduate
Professional Education as the Priority

C. Eugene Allen

At the end of the previous chapter, Joseph Kunsman discussed where we have been. I am a firm believer that we cannot do as well in planning for the future unless we reflect upon where we have been. For the area of agriculture, it is very appropriate to think about the canvas that Karl Brandt mentioned earlier in this volume. When I think of that canvas, I do not think of a bland one, but instead, I am reminded of the richness of the different eras of agriculture and how the mural depicts the changes that have occurred, from initially producing food for a small number of people at a local level to the numerous complexities of today's global dimensions for food and agriculture.

For example, in the beginning there was the labor-intensive era, when the majority of people worked to produce food for themselves and a few others. Then, there was the mechanically intensive era that led to the replacement of some labor by machines and increased the productivity of each producer. This was followed by a chemically intensive era that further reduced labor inputs, increased productivity, and was coupled with advances in transportation and preservation techniques that permitted food to be shipped longer distances. And as we modify that canvas today, those eras have not disappeared but they continue to evolve. The relative importance of these components to each other has changed; but we still have a mixture of labor, mechanical, and chemical inputs and now, increasingly, other inputs.

Perhaps what is not as clear as it was in past eras is what kind of era we are in today. That is part of what the conference was about. I would like to provide a reminder of some of the ways in which this era has been described, all of which are very important, since these topics relate to undergraduate education and to sustaining the agricultural, food, and natural resource system in the United States.

For example, this era has been described as an environmental decade, a biological era, an information and management-intensive age, and a global and international era; and it is an era when we must pay much more attention to cultural diversity. Finally, many believe that this is the decade for the undergraduate as it relates to program attention and curriculum content. Thus, the conference fit that description.

In each of the descriptions of this current era, age, or decade, there are major challenges and opportunities; and in each of these there is an explosion of information. So, is it any wonder that we are struggling with how to change and shape the curriculum to fit all of this together? As others asked a number of times, what do we give up in our programs and what do we choose to put in?

These are very difficult questions. As we think about the evolution of the mural that Karl Brandt described, we must include parts of all of the descriptions for the current era or decade. Thus, the depiction and integration of issues important in the last decade of this century are as difficult as the completion of any mural or picture.

We need to address some of the challenges with regard to specifics, namely, the curriculum, our goals for undergraduate education, and how we are going to go about delivering on these.

First, as has been described here, there are fewer people in the general population who know about agriculture. We also need to remind ourselves that more and more of our faculty come to our colleges knowing less about the breadth of agriculture. In some ways we are not unlike the Nobel Prize laureate in chemistry who did not know about plate tectonics, as described in Chapter 11 by Robert Hazen. How many of our faculty in agriculture could answer general questions about issues that are regularly addressed by the department next door?

We also have many students who come to us who have no previous connection to the food or agricultural system. That is very different than it was only a decade or two ago. Likewise, where students are employed when they graduate is very different, because we are working in an area where, in earlier decades, the primary focus was on production and marketing. Today, the balance has changed. Production and marketing are still important; but now the incorporation of new technologies, the need to access all kinds of data to make wise management decisions, and the need to understand and communicate this information to both professional and lay audiences are issues that we cannot escape. They are part of reality today, and the public and our students are demanding knowledge in these areas.

So, unfortunately, although we have fewer people, faculty, and students who know as much about the breadth of agriculture, today we have more demands on what is needed to deal with a wider array of complex issues.

Closely related to this is the second item that is part of our challenge, namely, that we must be more concerned about the environment, food safety, values, and ethics. These issues must not be peripheral to the substance of an agricultural curriculum but must be part of its core.

Third, there is an increasing sophistication among those producers, processors, and distributors who are carrying out these functions for the majority of the food in the United States. Their needs are increasingly sophisticated, systems oriented, and information based. All of these factors challenge the relevance of our teaching, research, and extension programs.

Fourth, we have a very serious human resource need, and this human resource need is not going to be addressed unless we do a much better job than we have done up to now by reaching out and encouraging minority populations to successfully complete educational programs in our colleges. Without more women and minorities in our educational programs, we will fail to capture the strengths from diversity and will fail to address the increasing need for talented human resources.

The fifth challenge is the international dimension. The international dimensions of food production and forest products are growing. They are becoming more complex, as reflected by the Uruguay Round of the General Agreement on Tariffs and Trade. We are also frequently reminded that there are people starving to death in many parts of the world. Such problems can be addressed only as teams of educated people with appropriate tools address the task of how to feed people living in areas where population pressures are frequently very severe and decisions on the use of land and water are extremely critical. All of these needs require an appropriate and culturally diverse education. It requires that those who work in such places be broadly educated and possess some specific knowledge appropriate to the priority needs. It also requires a much better understanding by people who set policy and who cannot avoid feeling the impacts brought about by circumstances in other parts of the world. In terms of the current situation in Eastern Europe, I raise the following simple question: As food moves into the international markets, what about the contamination of food grown and produced in places where the environment is so bad that the food is unsafe? Who is going to see where that food goes? Who is going to help correct those problems so that when that food is produced it is safe for consumption, regardless of where it is eaten?

The sixth challenge is one that a number of individuals addressed throughout the conference and in this volume. That is, how are we going to teach? Not just what are we going to teach but how are we going to teach? Are we going to continue with our old ways, or are there new ways that can bring about an improvement on the past? How will we use telecommunications to enrich the classroom and allow institutions to share the expertise of their faculty?

I also challenge us to think about the ideas Robert Hazen presented. If faculty in colleges of agriculture would choose to teach either applied or basic concepts in the manner described by Robert Hazen, how many students would be enrolled in our courses? I would venture to say that if we opened our minds to these concepts, we would frequently have to use our largest lecture rooms to accommodate many classes, because I believe that the majority of undergraduates would find that learning from such teachers would be a refreshing and rewarding experience. More faculty are going to have to more clearly differentiate the educational needs of the majority of undergraduates from those of graduate students and future scientists. We cannot do it all the same way. I fear that too many faculty view each group of students as potential future professors, and that is not where we should start.

We should all think about how we are going to deal with the topics that have been presented in this volume. These include the curricula of the departments or colleges, the courses and how they will be taught, our role in shaping the general and liberal education requirements of our universities, and the backgrounds and needs of our undergraduates. Other issues that we should consider relate to the role of disciplinary knowledge and its presentation in courses versus the interdisciplinary content and systems approaches; the use of both individual and team learning experiences; and how we can do a better job of addressing the issues of problem solving, communications skills, and ethics across the curriculum rather than expecting students to regurgitate facts.

Finally, what are our needs, the students' needs, and the needs of our students' employers? In this regard, what should we consider to be the requirements for bachelor's, master's, and even doctoral degrees? Isn't it rather sad that we turn out so many doctorates with excellent research skills but who are so ill prepared to communicate in lay language the significance of their research or so poorly prepared to teach an introductory-level class? What about the undergraduate who is required to learn the Kreb's cycle many times but who never comes to recognize its significance to energy metabolism? Or the student who learns about the role of many different nutrients but who does not begin to understand the complex of political, ethical, and natural resource issues that lead to death by starvation of millions of people virtually every year? Does such a curriculum seem appropriate? We need to ask questions and make changes to address such serious shortcomings. The future will be less forgiving to both those faculty and those institutions that fail to more adequately address such curriculum issues.

14

Science, Technology, and the Public

Peter Spotts

The focus of this chapter is science—one of the two significant trends that is shaping the future of agriculture. Robert Goodman explained this trend earlier in this volume. I begin with a high-tech tale of two cities: Cambridge, Massachusetts, and San Francisco, California.

In July 1976, controversy erupted in Cambridge over recombinant DNA research at Harvard and the Massachusetts Institute of Technology (MIT). This research was in its infancy then. In fact, on the day that public hearings were held during a Cambridge City Council meeting, the National Institutes of Health (NIH) released its safety guidelines for recombinant DNA research.

Some people—including Nobel Prize-winning physiologist George Wald of Harvard—argued that the two universities lacked sufficient safeguards to be allowed to conduct such research in the city itself. Nor did these opponents think much of the NIH guidelines. Both Wald and his wife, Harvard biologist Ruth Hubbard, argued that the NIH guidelines were self-serving and dangerously inadequate.

In contrast, other scientists argued that the research could be conducted safely under those NIH guidelines and that fears among the public and some of their colleagues about what could happen if genetically "tweaked" microbes managed to escape into the environment were overblown.

Clearly, something had to give. Recombinant DNA research held out the promise of significant advances in agriculture and medicine. Other industrialized countries were poised to compete with the United States in this field. Still, many people were concerned about the safety, if not the wisdom, of rearranging the fundamental building blocks of living organisms. The last thing Cambridge residents wanted was a strain of Teenage Mutant Ninja microbes running rampant up and down the sewers of Massachussetts Avenue.

The Cambridge City Council voted to establish a 6-month "good faith" moratorium on recombinant DNA research in Cambridge; Harvard and MIT halted their research. The city established the Cambridge Experimental Review Board, which was assigned the task of determining whether recombinant DNA work was too dangerous to be allowed within the city limits.

After 75 hours of hearings with testimony from all sides, 25 hours of discussions within the board itself, and still more hours spent wading through stacks of related documents, the eight-member board unanimously agreed that the research should continue, but only under a set of guidelines more strict than those of NIH. In addition, the city council was to set up a Cambridge Biohazards Committee, which would oversee all recombinant DNA research and report safety violations. It should be noted that not one of the eight members of the board that made these recommendations was a scientist.

How has the regime worked? I recently put that question to Robert Alberty, a chemistry professor at MIT who was dean of sciences during the period when the recombinant DNA debates in Cambridge threatened to split the city and the research community. He says that the regime is functioning very smoothly. The universities and the Cambridge Biohazards Committee are working very closely together.

Despite concerns that strict regulations might force researchers to move their work elsewhere, that has not been the case. In fact, he says, not only is the basic research progressing, but so is growth in the city's commercial biotechnology industry. At least a dozen firms now call Cambridge home. Why? Clearly, the presence of two prestigious universities—both leaders in the field of biotechnology—is a magnet. But ironically, says Alberty, so is the city's regulatory regime. Companies, he says, look at the tough regulations, see that they are workable, and set up their firms in Cambridge. It is better to locate where the battles have already been fought, they say, than to set up their firms somewhere that has yet to go through the same process.

Next, take the situation in San Francisco. In 1985, the University of California at San Francisco (UCSF) bought an office building in Laurel Heights, a mixed residential and commercial neighborhood. According to the residents, the university told them the building would be used for "academic" purposes, without being more specific. Initial opposition centered around issues of traffic and noise.

Opposition stiffened, however, when residents found out that the building would be used as a laboratory for 150 researchers from the school of pharmacy. Work would include research into parasitology, toxicology, and drug development. Community members worried about recombinant organisms, virulent germs, the venting of chemicals without treatment, and the use of commercial carriers for transporting radioactive isotopes to the facility.

Earlier in this volume, Frank Press mentioned the anecdote about the college-educated woman who stood up and said, "We know you're releasing DNA into the atmosphere, and we oppose it." The residents of that neighborhood were frightened. Fear is a very concrete emotion to those who experience it, even if it is based on a lack of knowledge.

The Laurel Heights Improvement Association sued the university, arguing that the environmental impact statement associated with the facility was inadequate. The association won, and the decision was upheld by an appellate court. The court accused the university of carrying out its activities in Laurel Heights "in a cavalier fashion." After the court shut down the laboratory, the university appealed to the state supreme court. In addition, the university announced that it would spend $1.6 million on environmental monitoring.

The state supreme court also held that the initial environmental impact statement was inadequate and sent it back to the university to be done again, although it also allowed work at the laboratory to continue. The Laurel Heights Improvement Association challenged the second environmental impact statement as well. In January 1991, a superior court judge tossed out the challenge.

The university is now proceeding with work at its laboratory, but at what cost? Apart from the cost of legal fees and two environmental impact statements, what can be said about relations with the community? Moreover, the original protests spilled over into other Bay Area universities, such as Stanford and the University of California at Berkeley. One Stanford media relations specialist called the UCSF experience "a public relations disaster."

What accounted for the difference in outcomes in these two examples? I suggest that the difference, in its broadest sense, can be summed up in four words: a sense of community. To my mind, when two universities openly discussed concerns about the nature and safety of scientific research and agreed to work with their host city in order to allay public fears, those universities displayed a sense of community that extended beyond that of a single discipline, category of disciplines, or institution. Although officials at UCSF may have thought that they were meeting community concerns, the results suggest that they were not. Remember, "cavalier" was the adjective the appellate court applied to the university in its initial handling of the laboratory issue.

When I peel back all the layers of the issues examined in this volume, I come away with a sense that, at its core, undergraduate education in science—be it agriculture or any other field—must help students know that they are part of a larger community, one that extends beyond the bounds of a particular discipline or even of the sciences as a whole.

I once spoke with a physics professor from MIT who marveled at the intelligence of his students. "They are absolutely brilliant," he exclaimed. "Social misfits, but brilliant." He may have been exaggerating for effect; but to me, his comment serves as a not-so-subtle reminder of the need to be sure that students—even in such specialized disciplines as are found in the sciences—emerge from their universities well-rounded and able to function in a society that looks upon "experts" with a mix of admiration and suspicion.

This is no mean feat. Often, it requires a way of thinking that in some ways differs from the scientific method. When starting a research project, a scientist may ask a "what if" question. Or he or she may be trying to test a hypothesis through experimentation. In either case, the approach is to conduct the experiment and see where the results lead. One may anticipate a result, or even a series of possible results. Ultimately, however, a scientist will await the outcome of the experiment before shaping his or her next move. Broadening the sense of and working with the larger community, however, requires anticipating and preparing for contingencies pegged to a range of possible outcomes.

It is becoming especially important to train the coming generation of scientists to consider the ethical, economic, environmental, and social effects of their work—to see that they have a broad, as well as deep, education.

The need for these skills in those who conduct and direct research is growing daily. If you need a reason, start with money. According to the National Science Foundation (1990), nearly half of the money spent for research and development (R&D) in the United States in 1991 came from the taxpayers. Their concerns, and those of the lawmakers whom they elect to represent them, are broader than how to pay for the next experiment or grant. It is to them, in magazine and newspaper articles, in television programs, and in testimony before legislative committees, that scientists must build a case for spending federal money on projects or defend current levels of federal R&D support. To millions of Americans, the need to spend several billion dollars on a high-energy physics facility buried deep in the heart of Texas is not self-evident.

Nor is it just the federal government that is signing the checks. The recession of the early 1980s put the fear of economic collapse into many so-called Rust Belt states. They began to invest in R&D efforts to help keep their economies going. Ohio's Edison program is one of the most frequently cited examples.

By 1988, according to a survey conducted by the state of Minnesota, 44 states spent a combined $550 million that year on programs to encourage R&D, foster innovation, and help boost the competitiveness of their industries. Programs range from Texas' support for high-temperature superconductor research and the superconducting supercollider to Utah's investment of $4 million for cold-fusion research. (That particular project is discussed in greater detail below.)

The $550 million is small compared with the amounts that the federal government and industry are spending. But given the amount of responsibility that the federal government has passed back to the states without giving them the wherewithal to follow through, that represents a noteworthy—and politically vulnerable—contribution to the country's R&D efforts.

All the more reason, then, to instill a clear sense of ethics and accountability in the future scientists and university administrators coming up through our institutions.

Unfortunately, the Bay Area and Boston more recently have illustrated what could turn out to be the less responsible side of handling federal R&D money. Recently, a congressional subcommittee called Stanford University to account for allegedly overcharging the federal government $200 million in indirect research costs. After a 5-month investigation, the General Accounting Office pointed to "serious deficiencies in Stanford's cost allocation and charging practices," as well as "inadequate oversight" by the Office of Naval Research, as the root cause of the alleged overcharges. Meanwhile, the General Accounting Office is looking into the Harvard Medical School's indirect research costs as well.

As for ethics in research itself, return to the example of cold fusion. The commercial use of fusion, by which two hydrogen nuclei are combined to release vast amounts of energy, has been pursued for decades. Unfortunately, the progress has been very slow, owing to the vast amounts of energy needed to fuse the nuclei and to uncertainties about the best way to confine the reaction. Indeed, one of my colleagues suggests that physicists have discovered a new physical constant: No matter what date you choose as a base, commercial fusion is always 20 years down the road.

Then, along came two chemists, one from the University of Utah and the other from Southampton University in Britain, who claimed that they had seen evidence of fusion at room temperature in a tabletop device. Researchers worldwide sought to replicate the results, with little, if any, luck. Subsequent analyses of their 1989 work appears to have uncovered a shift in a critical set of data that somehow occurred between the time the experiments were conducted and the time the two published their work—with no explanation for the change.

One physicist, Frank Close of the Oak Ridge National Laboratory, and the Rutherford Laboratory in Britain claim that the two chemists violated scientific ethics. Another, Richard Petrasso of MIT, says that he has downgraded his criticism from outright fraud to a violation of how science should be done. The researchers themselves stand by the validity of their work.

Where is the harm in this? Apart from the hopes that were raised and then dashed, there is the issue of $4 million that the state decided to contribute to cold-fusion research. Four million dollars may not sound like much, but it is money that the state might have spent on its public schools, for example, where not long ago teachers were complaining of poor salaries and overcrowded classes. Representative Ray Thornton warned earlier in this volume that scientific illiteracy may result in misguided public policy and misdirected public money; so, too, can badly or fraudulently conducted research.

Aside from the ethical questions related to the conduct of research or the oversight of research funds, there is the issue of the ethics involved in pursuing a particular line of research—and the ability to communicate them to a wider group. I recall an interview I had with Thomas Wagner, who, at the time, was the director of the Edison Center for the Study of Animal Biotechnology at the University of Ohio. The discussion was broadly centered on a U.S. Patent Office ruling that genetically engineered animals—except *Homo sapiens*—could be patented. The ruling sent activist Jeremy Rifkin and others through the roof. They were dismayed at what they termed the mind-boggling moral and ethical issues raised by the Patent Office's ruling.

The ethical question was succinctly posed by Philip Bereano, an associate professor at the University of Washington who taught courses in technology and public policy. "Where do we draw the line?," he asked. "Do humans exist in and with nature, or is nature for man's exclusive use?" (Spotts, 1987b:1).

When I turned the questions to him, Thomas Wagner replied, "I put animal life forms into three categories philosophically and morally. The first is wild animals who live in a natural ecosystem; the second is agricultural animals living in a synthetic ecosystem; and the third is human beings. I think the patent decision speaks to the second category. Since we have created an artificial ecosystem, we almost have a moral requirement to alter farm animals to fit that ecosystem." He added that there's "no way that man can make a deer better able to survive in the wild than nature" (Spotts, 1987b:1).

While one may or may not agree with his line of reasoning, I found myself appreciating the fact that he had clearly thought through the ethical or moral component of his work and could clearly state it to someone who is outside the field.

This brings up another point; that is, as educators consider ways to train the next generations of researchers, they must pay equal attention to instilling in the rest of their students a level of scientific literacy that helps them to respond intelligently when public policy issues affecting science and research arise. It is interesting to note that when Stanford University was feeling the ripple effects of the donnybrook over UCSF's Laurel Heights research laboratory, Stanford president Donald Kennedy argued that scientists and educators were partly to blame for a "disappointing level of scientific literacy" that he felt underlay the dispute.

The late Roger Nichols, a microbiologist who left Harvard Medical School after a productive career in research and teaching to become head of the Boston Museum of Science, put it to me this way by drawing an analogy to the Middle Ages:

> [A]rguably the preeminent cultural trait was religion and the preeminent cultural institution was the church. Yet unless we spoke Latin, or read or wrote Latin . . . we would have had to sit there nodding our heads when the priests said this is good for you, because we could not participate in the theological discussion.
>
> That's where we are today. The preeminent cultural traits of our time are science and technology. And yet most of our people are disenfranchised from participating in the preeminent cultural traits of our time. . . . The echoes of the Reformation which was really people saying, 'We no longer want our major preeminent cultural trait to be conducted in a language we can't understand'—are still with us (Spotts, 1987a:3).

As educators consider training the priests, they also should consider training the laity—giving them enough of the language to allow them to ask basic questions.

How might one do this? Instead of teaching introductory biology straight from a textbook, why not base the course on an issue, such as the open-air testing of genetically altered microbes. One still would have to deal with the basic science. But for someone not going into research, it becomes increasingly clear that this is a subject he or she should pay attention to, because it is an issue that is taken up at the most fundamental levels of government, the city or town. Remember that one of the key battlefields over the open-air testing of frost-retarding bacteria in 1987 was the hearing room of the Monterey County (California) Board of Supervisors.

For the student who intends to go on to a career in research, such an approach to teaching would communicate clearly from the outset that ethical, economic, and environmental issues—and the ability to discuss them—are as crucial to the progress of his or her career as a grasp of the basic science involved is.

The role of educators, however, does not stop at the classroom door. Whether educators are comfortable with the prospect or not, the fact remains that they also educate by their responses to questions posed by members of the media. They not only educate the reporter, but if they are doing their job right, they are helping to educate the public as well. I am not speaking of public relations here, at least in the narrow sense. To me, public relations is what someone engages in after the reactor has melted down.

Taken in a broader context, however, maybe the phrase fits. Many reporters may come to scientists with little background on the subject they want to pursue. They should be treated the same way a freshman student would be treated. The most gratifying experiences I had as a science writer were those spent with researchers who were willing to stick with me until it was clear I understood what they were talking about. Each story was the fruit of miniseminars—whether on some discovery in a particular field or on some public policy question.

I close with a plea. Our oldest child is in the first grade. He obtained an honorable mention in his elementary school's science fair. Will he sustain his interest in science? I hope so. Will he pursue a career in science? Perhaps, but that is his choice, not mine. Whatever the answers to those questions, I hope that scientists and researchers and those they train will include in their sense of community my children and millions of others—and their parents. Through whatever outlet available and with all the patience that can be mustered, help them learn the language of the priesthood.

References

National Science Foundation. 1990. National Patterns of R&D Resources 1990. NSF Report 9316. Washington, D.C.: National Science Foundation.

Spotts, P. 1987a. Science does not have to be 'all Latin' to adults, says museum director. Christian Science Monitor, March 24, 1987, p. 3.

Spotts, P. 1987b. U.S. stands at cross-road on genetic alteration. Christian Science Monitor, April 27, 1987, p. 1.

15

A Challenge, a Charge, and a Commitment

Karl G. Brandt

I would like to return to the metaphor of the blank canvas that I used earlier in this volume (see Chapter 1). Is your image for a new curriculum beginning to emerge? I admit that mine still has a few problems. It is still an unfinished piece, but it does have new shapes and colors. Although it is not ready to be hung in a gallery, it is different from what I brought with me. I hope that the conference participants and the readers of this volume have something exciting on their canvases.

In a more serious vein, I would like to point out a parallel. During the conference, a selection of books was on display. One of them was a tan paperback volume published by the Board on Agriculture of the National Research Council, *Investing in Research* (National Research Council, 1989). The idea behind that publication was the generation of substantial new support for the research and development activities that are major efforts on all of our campuses.

The parallel is another document, one that was found in the folders of the conference participants, a thin, rose-brown conference program brochure entitled "Investing in the Future: Professional Education for the Undergraduate." By cutting and splicing, that title can be changed to read simply "Investing in Education for the Undergraduate."

The colors of these two documents are, by and large, the same. Both publications carry the imprimatur of the Board on Agriculture of the National Research Council. Both have something important to say about what goes on in colleges of agriculture. There is a problem, however. *Investing in Research* is much thicker than "Investing in Education for the Undergraduate."

I hope the papers in this volume have generated many good ideas that, when they are implemented back on our nation's campuses and moved toward fruition, will serve as subjects for proposals to the Higher Education Challenge Grants program of the U.S. Department of Agriculture. Good ideas that are well expressed, whether they be ideas for research or ideas for better educating the young people of our country, do get funded. I hope that your ideas will attract funding to strengthen and support the education of students in our colleges and universities. In so doing, your creative energies will provide the wherewithal for expanding the size of "Investing in Education for the Undergraduate," so that, in weight and accomplishment, it grows to match *Investing in Research*.

Education and research are vital—they are parallel tracks on which ride the processes of human capital development and scientific discovery that fuel the engine of our industry, but they must be tracks of more nearly equal load-bearing capacity if we are to move forward with confidence. If one is weaker than the other, derailments are inevitable. We have a way to go, but the conference and this volume point us in the right direction.

In closing, I leave you with this. Conference participants and readers of this volume have been encouraged to start with a blank canvas and create a new vision—a picture of a new curriculum for educating the students in their colleges of agriculture. Pictures are nice, and they can inspire. But if they only hang in the gallery on your campus and are never made real, they will only be reminders of what might have been.

The challenge to each participant and reader is to turn art into life. Accomplishing such change is a very human activity and a very political activity, but it is essential. Will it happen on your campus? It is up to you.

Reference

National Research Council. 1989. Investing in Research: A Proposal to Strengthen the Agricultural, Food, and Environmental System. Washington, D.C.: National Academy Press.

Part II
Conference Discussions

CHAPTER

16

Teaching and Research:
Balance as an Imperative

Anne M.K. Vidaver and Arthur Kelman

Francille M. Firebaugh, First Rapporteur
Mort H. Neufville, Second Rapporteur

A provocateur's objective is to stimulate thought, discussion, and debate. This can be achieved by proposing both conventional and unconventional approaches to the task of balancing teaching and research. Such a balance is desirable at both the undergraduate and the graduate levels.

Where Are We Now?

Concern for undergraduate education in the United States is much in the news these days. This concern covers the spectrum of educational institutions—from liberal arts colleges to land-grant universities—and is increasingly expressed as part of a larger continuum of concern about our entire educational system. Thus, the deficiencies of our educational system have been identified—perhaps in the starkest terms—by the data indicating how low the test scores of our students are relative to those of students in other industrialized countries. Notwithstanding the recognition that such tests do not necessarily measure many traits that we hold dear, such as creativity, perseverance, and the ability to synthesize ideas, there is general agreement that the nation has a serious problem, especially in the sciences (Boyer, 1990). Students' lack of knowledge about the importance and nature of agriculture, forestry, and natural resources is an even greater problem. The search for resolutions to these problems was the mission of the conference on which this volume is based, from balancing teaching and research to breaking traditions in curriculum design. The conference sought innovative ways to attract students to the areas of agriculture and natural resources and to involve colleges of agriculture in educating an increasingly urbanized citizenry about agriculture. The addition of the term *natural resources* to agriculture reflects the fact that both managed and natural ecosystems are valid areas of study and research in colleges of agriculture, even if they are not always so recognized.

Factors That Adversely Affect Emphasis on Teaching

Before suggesting some steps that can be taken to enhance the balance between teaching and research, it is important to examine the factors that adversely affect teachers. First, there are many forces at universities that serve to divert faculty from making the commitment that is necessary to motivate students in the learning process. The pressure on faculty to obtain grants from the federal government and private industry has increased in recent years. In part, this reflects the fact that base support at most institutions has been seriously eroded. In most grant programs that are open to research scientists, the competition has intensified as the number of scientists seeking funding has increased. In the interval between 1977 and 1987 there was a 60 percent increase in the number of research scientists at universities in the United States (Abelson, 1991). Although funding for research increased, it did not keep pace with the increase in scientists and the increases for overhead and the basic costs for supplies and equipment. As a result, in relation to the number of grant applications received, the percentage of grants, particularly in the biological sciences, funded by the National Science Foundation and by the Competitive Grants Program of the U.S. Department of Agriculture (USDA) has declined. It is currently between 15 and 20 percent.

Of particular concern for young faculty at the start of a tenure-track appointment is the need to obtain an initial grant. Often, at this stage, new faculty are assigned the responsibility for teaching the introductory course in their field. The broad background needed to teach such courses demands a major commitment of time. When tenure is at stake, it is not surprising that the tilt in allocation of time will be toward research and the preparation of grants at the expense of teaching commitments. Although there is a tendency not to acknowledge this impediment to teaching, it is necessary to recognize the fact that the financial and professional rewards for excellence as a teacher are rarely equivalent to the rewards for outstanding contributions in research (Sykes, 1988).

In 1989, the Carnegie Foundation for the Advancement of Teaching completed a survey of faculty at research, doctorate-granting, comprehensive, liberal arts, and 2-year institutions (Boyer, 1990). The survey presented a number of questions concerned with attitudes toward research and teaching assignments. Several of the questions related closely to the subject of balance and the recognition of teaching versus research. The following statements and responses were obtained from faculty at doctorate-granting institutions, which would include all the research-oriented land-grant universities with colleges of agriculture:

1. In my department it is difficult for a person to achieve tenure if he or she does not publish. Seventy-one percent strongly agreed and 18 percent agreed with reservations.

2. At my institution publications used for tenure and promotion are counted but not measured quantitatively. Fifty-three percent agreed.

3. At my institution we need better ways besides publications to evaluate the scholarly performance of the faculty. Seventy-seven percent agreed.

4. During the past 2 to 3 years, financial support for work in my discipline has become harder to obtain. Sixty percent agreed.

It is apparent from these responses that most faculty believe that research should be their first priority. In a survey of the chief academic officers at doctorate-granting institutions, the question was raised as to whether in the evaluation of faculty performance the balance in teaching, research, and service has shifted. Fifty-six percent of the respondents agreed that there had been a significant shift, with increased emphasis on research at the expense of teaching and service. These surveys provide additional evidence to support the perception that outstanding performance as a teacher is not rewarded to the same degree as an outstanding contribution in research is, notwithstanding the efforts to redress this situation.

A second major factor that can affect teaching adversely is the service requirement, including extension-related activities, that can impinge directly on the time available to prepare for teaching courses. Most faculty involved in teaching undergraduates in colleges of agriculture have appointments that combine teaching with research and service commitments. There is constant pressure to respond to requests for advice from growers, industry representatives, and county and extension colleagues. Service, in the broad sense, also includes the need for faculty to be involved in faculty governance, with commitments to departmental responsibilities as well as college and university assignments. Faculty are also expected to review papers and grants for colleagues and to serve as panel members for various granting agencies. In addition, as members of professional societies there is the expectation of service on committees, editorial boards, and related society functions. Senior faculty members are often asked to serve on national advisory boards or committees of government and private foundations.

A third major commitment of time is associated with service as a mentor for young colleagues and faculty in other disciplines and involvement in advising undergraduate and graduate students. When one considers the other demands on a faculty member's time, it is necessary to decide not only how to balance teaching and research but how to balance these two activities with the service component as well.

Another major effect on motivation for teaching is the lack of positive feedback, particularly for an individual at the start of his or her career. Excellence in teaching is rarely recognized until a teacher has taught for several years and students begin to spread the word that an individual teacher is outstanding. Once recognition comes via awards and other evidence of faculty support, such recognition is usually within a college or department and rarely extends to the national or international level. In contrast, a breakthrough in a specific research area may result in very rapid national and international recognition and professional advancement. More recently has arisen the prospect that a major advance in research can be patented and royalties may be forthcoming, particularly in the area of biotechnology. Financial rewards also can come from consultantships and increased research funding.

An additional negative impact on teaching is the fact that there is always a certain degree of ambiguity in defining the responsibilities of any faculty appointment. It often is not clear how much time a faculty member, especially a new member, should invest in teaching at the expense of research and service. When one considers all of the above factors that impinge negatively on teaching, it is perhaps surprising how many instructors are still willing to devote the time, energy, and thought needed to be outstanding teachers.

Steps to Improve Balance

Outlined below are some steps that can be taken to enhance the balance between teaching and research.

1. *Reward both teaching and research.* To ensure that teaching and research are seen as both essential and complementary activities of all faculty, both must be rewarded if both are to be performed well. The impetus for reward must be implemented from above, and the promotion and tenure system, which is at the heart of the reward system in academe, must reflect this equity of emphasis. Such concerns have emerged in various forums (Koshland, 1991; Pelczar, 1990).

2. *Minimize teaching of "how-to" courses.* A debatable proposition to balance time in teaching and research is actually to decrease the teaching of training or "how-to" courses to a minimum. This recommendation is made because the pace of modern science continues to accelerate, which makes it even more difficult for university and college professors to keep up with their respective disciplines and take time to teach (Koshland, 1991). As noted above, added to this time crunch is a necessity to secure funds to support the research for which professors are employed.

3. *Panel teaching should be explored.* By *panel teaching*, we mean that there should be courses that are taught simultaneously—not a sequence of individual lectures by a number of professors from the same department but lectures by several instructors from different fields. Such courses would emphasize the integration of ideas, consider differences in perspectives, consider how different disciplines interact, and simultaneously provide a forum for peer evaluation of teaching. This suggestion is proposed because it is curious that in real life, people such as farmers, small business people, producers of value-added products, or persons involved in global commodity trading integrate multiple inputs of information in their decision-making processes. However, students rarely observe such integration in college and universities. It is principally in sports that a team approach by the coach (teacher) and players (students) is rewarded in universities. Under a team system, all participants strive to reach a common goal. Interdisciplinary research by faculty, a form of integrated problem solving, is being tangibly rewarded in some universities, but students rarely witness such activity.

For example, in the area of crop protection or sustainable agriculture, it would be valuable to have students exposed to a problem from the merged perspectives of the plant breeder, entomologist, plant pathologist, soil scientist, agricultural meteorologist, weed scientist, ecologist, social scientist, and/or biochemist working as teaching teams or panels, all of whom should be familiar with the latest thinking and developments in their respective fields. Students should be exposed to the process of integrating and synthesizing knowledge based on different viewpoints and should not have to learn to integrate knowledge solely through postgraduate experience. Such panels would also serve the dual purpose of encouraging research and encouraging teaching coordination and communication among faculty. Such a panel approach could affect the total teaching time of faculty, either positively or negatively. Faculty need increased time for the assimilation and synthesis of knowledge; the amount of new information is overwhelming. The panel approach for certain advanced courses ideally could alleviate that constraint.

In many situations, such a panel approach has proven to be highly effective. For example, the longest-running program on public television in Nebraska, *Backyard Farmer*, has had a panel format throughout its 39-year history.

4. *Specify teaching roles for adjunct faculty.* Each adjunct faculty member should have a specified teaching role. Many campuses are fortunate, for example, to have USDA personnel who may supervise graduate students in research. However, it is unfortunate that these scientists must teach on their own time, if they so desire and if their own and affiliated administrations approve (we know of only one exception). Cooperative agreements nationwide should be modified to allow for a teaching role for USDA and other scientists, and if congressional approval is required for such action, it should be sought. Such a role for adjunct faculty will lead to increased professionalism and improve the prospects for U.S. competitiveness and the future education and training of students in the sciences and technology. Thus, the teaching burden on state-funded faculty would also be lessened.

5. *Teaching schedules should be flexible.* Teaching should be flexible with respect to the times and credit offered. Such flexibility could permit a reduction in a particular faculty member's teaching time by fewer credit hour requirements or at least offer more efficient use of time. All the courses in colleges of agriculture seem to be tied to a standard daytime regimen. Where are the weekend or evening courses that might be attractive to instructors and that might also attract students? As we vie for nontraditional students, courses need to be presented at nontraditional times. Also, minicourses should be identified, and courses should be presented as short modules. Both individual and panel instruction could potentially equalize teaching loads and would be adapted to the needs of the clientele—students, many of whom work part- or full-time. This could improve science literacy, one of the goals for the year 2000 of the Federal Coordinating Council for Science, Engineering and Technology's Committee on Education and Human Resources.

6. *Students should participate in teaching.* In at least some advanced courses, students should participate in teaching, with the instructor as guide or mentor. This suggestion could potentially lessen the burden of preparation time for faculty. Such participation occurs in some graduate courses. This may rekindle the enthusiasm of undergraduates in some cases, a characteristic that is considered by some critics as being in short supply (Sanoff, 1990).

7. *Learn to value good teaching.* Universities may need to learn from other sources how to value good teaching and aim for more coordination between subjects and disciplines in teaching. There seems to be a general opinion, justified or not, that small liberal arts colleges are doing something right. Furthermore, projects such as the Department of Education Fund for the Improvement of Post-Secondary Education can serve to improve teaching at research universities. At the University of Nebraska, for example, this in-

volves critical self-evaluation by faculty, with administrative support, that leads to specific rewards for effective teaching (Narveson, 1990). Still missing from all materials and analyses, however, is what we can learn from the industrialized nations that have a better record in science and technology education than we do. What can be used or modified that would be useful to the U.S. educational system?

8. *Improve the ways to prepare teachers.* We need to improve the ways in which we prepare graduate students and young faculty for their roles as teachers. Many innovative approaches have been suggested recently (Barinaga, 1990; Lee, 1990; Palmer, 1990; Pool, 1991). In addition, good teachers recognize that effective teaching requires application of basic well-defined skills in communication. There are very effective guidelines for success in communicating ideas in the classroom and in evaluating success in teaching by testing students for their knowledge and comprehension of the subject matter as well as ability in synthesis, application, analysis, and evaluation (Bloom, 1956). All instructors can gain from personal contact with the master teachers in a department. This type of mentor supervision is often lacking and may be preferable to requirements for a short course in teaching methods for those individuals who have never had the opportunity to benefit from supervised guidance in a teaching position.

9. *Revise the current concept of scholarship.* In his excellent treatise entitled *Scholarship Reconsidered: Priorities of the Professoriate,* Boyer (1990) emphasized the need to revise our current concept of scholarship that places research as the preeminent activity of a university professor. Boyer advocates a broad definition of scholarship that would place teaching as well as integration and application of knowledge on the same plane as research or discovery of new knowledge. He also outlines in some detail how this can be accomplished.

The question is often raised as to whether an active research program enhances the ability of a faculty member to teach in a more competent and effective manner than if he or she had no research experience. Most faculty would agree that the opportunity to engage in research does enhance the quality of teaching. However, it is often not evident to young faculty members that the reverse may be true.

Needs for Coordination and Communication

In balancing research and teaching, one of the most difficult challenges to solving problems and generating new knowledge in this age of complexities is the need for increased coordination and communication among faculty. This leads to both commendation and criticism from the Alliance for Undergraduate Education, encompassing 16 major universities, whose main goal is "to foster major innovations in the way undergraduate science is taught, and to forge more intimate links between research and teaching" (Certain, 1990:2). The alliance, which has the role of leadership in science and engineering education, has not clearly included agriculture within its purview. The alliance has identified areas of concern and produced an agenda for revitalizing freshman-level courses. Some of these objectives are particularly applicable to agriculture because of its historical development apart from the other sciences dating back to the nineteenth century. The alliance recognizes the importance of collaboration among the different disciplines, as does the National Science Foundation: "Future agricultural scientists will need skills and knowledge outside traditional agricultural disciplines" (National Science Foundation, 1989:4).

Furthermore, we must find a way to involve colleagues in industry in the education and training of both faculty and students. There is something fundamentally wrong in our educational system when senior officials in industry claim that, with the possible exception of biotechnology, innovation and incremental advances in knowledge most often occur in industry (National Academy Press, 1991). We need candid assessments of how to change the conditions that foster this perspective, if we consider that this is incorrect, as well as how to improve the training aspects of both faculty and students to solve the problem.

In some areas of agriculture, such as biotechnology, we should take to heart the recommendation of the National Research Council report that says "highest priority should go to increasing the retraining opportunities available to university faculty and federal scientists to update their background knowledge and provide them with laboratory experience" (National Research Council, 1987:107).

Strategies for Management of Time

In addition to the approaches discussed above, it is helpful to consider how individual faculty members can manage time more effectively in their search for balance. It is essential to recognize the critical importance of time management when dealing with competing demands. The concept that needs to be emphasized is the necessity of recognizing that one must "invest" his or her time rather than "spend" it on daily assignments. The ability to organize one's time is the most precious of all skills. Priorities need to be established on a daily basis; this enables one to avoid the so-called drippy faucet syndrome, in which attention is given to minor doable assignments in preference to dealing with major or long-term projects that warrant priority for completion. Many faculty would profit from a short workshop on time management.

Conclusion

There is much that can be changed. Although this chapter has dealt principally with faculty issues, consideration also needs to be given to what balance means to students relative to how they learn, what they learn, and their programs of study. We have suggested a few possibilities that can add to the consideration of how to improve undergraduate education, including how to balance teaching with research and service.

References

Abelson, P. H. 1991. Federally funded research. Science 252:1765.

Barinaga, M. 1990. Bottom-up revolution in science teaching. Science 249:978–979.

Bloom, B. S., ed. 1956. Taxonomy of Educational Objectives: The Classification of Educational Goals. Handbook 1: Cognitive Domain. New York: David McKay.

Boyer, E. L. 1990. Scholarship Reconsidered: Priorities of the Professoriate. Princeton, N.J.: The Carnegie Foundation for the Advancement of Teaching.

Certain, P. R. 1990. The Freshman Year in Science and Engineering. A brief report of the Alliance for Undergraduate Education, University of Wisconsin, Madison.

Koshland, D. 1991. Teaching and research. Science 251:249.

Lee, M. W. 1990. Turning teachers on to science. Science 249:979.

Narveson, R. 1990. From regard to reward: Improved teaching at a research-oriented university. Teaching at UNL 11(4):1–4.

National Academy Press. 1991. Industrial Perspectives on Innovation and Interactions with Universities. Summaries of Interviews with Senior Industrial Officials. Government-University-Industry Research Roundtable and Industrial Research Institute. Washington, D.C.: National Academy Press.

National Research Council. 1987. Agricultural Biotechnology: Strategies for National Competitiveness. Washington, D.C.: National Academy Press.

National Science Foundation. 1989. Profiles—Agricultural Sciences: Human Resources and Funding. NSF Report No. 89:319. Washington, D.C.: National Science Foundation.

Palmer, P. J. 1990. Good teaching: A matter of living the mystery. Change 22:10–16.

Pelczar, M. J., Jr. 1990. Microbiology education: The issue of balance. American Society for Microbiology News 56:516–517.

Pool, R. 1991. Science literacy: The enemy is us. Science 251:266–267.

Sanoff, A. P. 1990. The university in chaos. U.S. News & World Report, May 7, 1990, p. 16.

Sykes, C. J. 1988. Profscam: Professors and the Demise of Higher Education. Washington, D.C.: Regnery Gateway.

FIRST RAPPORTEUR'S SUMMARY

Throughout Anne M. K. Vidaver's and Arthur Kelman's opening comments, observations and generalizations about agriculture were made. They were often made as statements without rebuttal or discussion, so they should be taken in that light.

The definition of the domain of agriculture is critical to determining the balance of what is taught and researched and the future of colleges of agriculture. Much of the discussion about agriculture concerns its maintenance as a field, without addressing the basic issue of the domain of agriculture. Science is rapidly moving ahead of many colleges of agriculture; one participant felt that only by eliminating most current departments would the needed radical changes in colleges occur.

Agriculture will have fewer resources in the future, and colleges must downsize. They must also develop new approaches and attract new students. To remain viable, colleges will have to change. Professional societies create barriers to change through accreditation requirements that more often reflect current status rather than future directions. The historical base of organization of professional groups and their lack of response to change may not impede changes in colleges of agriculture, but it may also not contribute to a climate for change.

Agriculture needs to be concerned about minority issues and to encourage minority students to study science. Although some advances in gender balance have been made within colleges of agriculture, the enrollment of more undergraduate and graduate minority students and their graduation are critical to achieving diversity.

Instruction

The need to broaden course content and to make courses of greater interest to more students was expressed by several participants. More courses that are not strictly discipline based should be available; "the problem is finding courses to teach," reported one participant. It was suggested that a basic conflict exists in hiring discipline-based scientists with a focus on production agriculture problems and in expecting them to teach broad-based interdisciplinary courses. Agriculture should capitalize on the current interest in the environment.

Flexibility in course schedules, including weekend and minicourses, should be increased to better serve nontraditional students. The possibility of teaching a course by a panel of several instructors from different fields should be explored, emphasizing integration and the interaction of different disciplines and giving students the advantage of observing differences in faculty perspectives.

The importance of having the best teachers offering introductory courses was stressed. Courses that include writing requirements and regular feedback on papers and exams as well as cohesive capstone courses were described as time intensive in nature but as having a great potential to better teach and integrate course content. Vidaver and Kelman urged that teaching of training or "how-to" courses should be kept to a minimum; others commented on the benefits of providing students with hands-on experiences. They felt we should better relate courses to the needs of students.

Teaching and Rewards for Teaching

The faculty reward system should (1) be consistent and based on clear expectations, (2) reflect the assessment of both teaching and research as complementary and essential activities, and (3) provide incentives for teaching excellence and curriculum innovation. Promotion and tenure guidelines may actually be disincentives for teaching. Guidelines that specifically engender a greater emphasis on teaching are needed. It was noted that teaching evaluation should be at the level of research evaluation and that we need more peer review of teaching and review of course examinations.

A suggestion that did not receive general support was to require and reward excellence in teaching before requiring and rewarding excellence in research. A recognized problem is that new faculty (indeed, many faculty) reduce their teaching involvement by "buying out" of courses to increase the time they can devote to research. It was attributed directly to the promotion and tenure system of rewards.

Recognition should be given to the importance of teaching, including the option of devoting sabbatical leaves to developing innovative teaching methods and courses. Structured programs for the improvement of teaching should be encouraged, and adequate support should be made available. Recognition and rewards for attracting additional students in courses should be given to faculty in areas where there are declining enrollments.

Vidaver and Kelman indicated that it would be advantageous to have adjunct faculty, for example, U.S. Department of Agriculture personnel, with teaching responsibilities. No response to the idea was made. The merit of assigning faculty to teach areas closely aligned with their research focus and methodologies was stressed.

The need for teachers who themselves have the ability to assimilate and synthesize ideas and who challenge students to develop these skills was recognized. We need to integrate undergraduates more fully into the department and disciplines through participation in research, in teaching, and through opportunities for social exchanges with faculty. Students who participate in teaching, with the professor serving as a mentor for the student teacher, increase their identification with the goals of the department and the discipline.

Pleas were made for coordination and communication between and among such groups as the Alliance for Undergraduate Education, colleagues in industry, and other university faculty with faculty of colleges of agriculture. One participant noted the potential for faculty from colleges of agriculture to learn from faculty in U.S. liberal arts colleges and from other industrialized nations about good ways to teach science and technology.

SECOND RAPPORTEUR'S SUMMARY

Anne M. K. Vidaver and Arthur Kelman presented an overview of the basic issues confronting higher education in agriculture as administrators and faculty attempt to create and ensure a balance between teaching and research. The paramount question is how to ensure that the two are complementary and not divergent.

In attempting to address the question, Vidaver and Kelman outlined four issues that are basic to the resolution of the problem.

Factors That Influence the Quality of Teaching

There are pressures of competing forces—teaching versus research versus service. We sometimes neglect the service function, particularly the on-campus service function. The pressures result from a competition for time, including time for university governance, professional activities, and mentoring. The mentoring aspect involves a tremendous time commitment, not only for students but also for young faculty, and it is often overlooked.

Creativity and effectiveness are often difficult to evaluate in teachers. The rewards for good teaching are never equivalent to those for excellence in research. We are the problem in believing that there is a problem in evaluating or rewarding teachers. No one prevents us from effectively doing it.

Reference was made to the possibility that teachers hinder the creativity of students. Outstanding students prefer to be challenged and not stymied. The following question was also raised: Are we prepared for the challenge of teaching a class of students with diverse backgrounds? In addition, doctorate-level students who teach sometimes lack training in the art of communications, for example, techniques in managing a class and the actual teaching process. We have a responsibility to mentor our peers and create internal structures to improve and reward teaching so that faculty can effectively respond to these challenges.

Factors That Favor
an Emphasis on Research

We tend to instill in our students the idea that the path to gaining preeminence is research rather than teaching. This is somewhat related to the ready accessibility of tools for measuring creative and effective research.

The mentoring rewards are evident in research endeavors and through funding, publications, and graduate assistantships; but it is not evident in the teaching program, where there is a pattern of delayed recognition and a lack of positive feedback. It was pointed out that there is an ambiguity of expectations.

Faculty paid exclusively out of the teaching budget are also expected to consult and do research. Researchers who are paid exclusively out of the research budget, however, are not required to teach.

Strategies to Enhance Balance

Is involvement in research essential for effective teaching? The response was that more of an impact is made on students' lives through their involvement in active learning experiences. For example, the experiences of a student working alongside a bench scientist can be an effective stimulus to creativity and knowledge assimilation.

Teaching and research are integrative processes that contribute to the genuine overall education of students and teachers. There are grants in the teaching profession, and there are grants in research. We should not track students on the basis of their interest but, rather, try to truly educate them and not create clones after our own image. The numbers game is rather pervasive on our attitude toward teaching and research: We generate so many publications and teach so many students without any reference to the effectiveness of our teaching or our research endeavors. The pressure to publish is being driven by the tenure system, and all educators are a part of that system. This must change so that equal recognition can be given to excellence in teaching.

During the discussion, individuals were asked to respond to the issue of enhancing balance from the perspectives of a dean, a department head, and an assistant professor.

Dean's Perspective

• Strong administrative support for the integration of teaching, research, and service is the original spirit and mandate of the land-grant university philosophy.

- There must be advocacy for and commitment to excellence in teaching as well as research and service and a strong emphasis on quality among students, faculty, and support staff.
- A comprehensive program of teaching awards, enhancement grants, salary enrichment, faculty teaching development programs, and recognitions should be emphasized.
- There should be a strong and challenging undergraduate research and scholarship program.
- One of the problems associated with teaching is that it is done and evaluated locally, which is not true of research and, to an even lesser extent, extension. This leads to issues related, first, to how evaluation is done locally and, second, the difficulty for a teacher to become established nationally among his or her peers. Many faculty respond to these forces by deciding to emphasize research. This situation needs to be addressed by good local evaluation and reward systems for teaching. For example, at the end of the year, when raises are given, there should be clarity on the impact of the contributions or lack of contributions in teaching, research, or extension on that raise. Too many raises are attributed to the research efforts of the faculty when, in fact, it may have been due to the teaching efforts.
- Prospective faculty members should not only give a research seminar but should also give an introductory class lecture and be seriously evaluated on both during the interview process. This sends a signal to everyone in the department that teaching and research are important.
- Evaluations of each area must be objective; that is, evaluators should be wary of student evaluations. Performance criteria should be set at the beginning of the evaluation period.
- In cases in which split teaching appointments are in place (two or three way), the evaluation should be balanced accordingly.
- Evaluations should be ongoing, not once per year.
- Teachers must become directly involved in creative activity (not necessarily original research). New ways to teach information should be developed. Researchers must be involved in teaching at some level.
- Organizational philosophy must be clearly delineated by the dean.
- There should be encouragement and support for good teaching and/or good research (scholarship).
- The nonpriority service responsibilities of junior faculty should be minimized.
- Departments should be provided with the funds to do the job, not to micromanage full-time equivalents, for example, a 40-40-20 split.
- There should be uniform teaching, research, and service assignments for all faculty in a department. Teaching, research, and extension should not be micromanaged.

• Teaching should be built on a variety of approaches: advisement of individual students with individual faculty, mentoring, individual research assignments, laboratory efforts, field trips, summer internship experiences, classroom teaching, student clubs, and guest lecturers from industry, government, etc.

Department Head's Perspective

• When the undergraduate is integrated into the research enterprise, excellence in teaching and research will tend to be parallel.
• There is no fundamental premise that deficiencies in undergraduate education occur because professors are too busy doing research.
• Student evaluation of teaching is an effective tool both for the evaluation of teaching and the modification of techniques. It should be supplemented by collegial evaluation.
• There is a richer diversity of teaching situations than just classrooms. A good teacher who researches in an academic setting is always teaching by example. We should broaden the public's perception of what teaching is.
• The academic senate (faculty) should have a powerful voice in the academic personnel process.
• The idea that time is best spent in the laboratory should not be promoted. Rather, balanced programs should be developed.
• Faculty should be encouraged to share research findings (including mistakes and problems) with their classes.
• Special-problem courses should be developed so that undergraduates can be exposed to research techniques, concepts, and attitudes.

Assistant Professor's Perspective

• A young assistant professor should be able to choose two of the categories on which he or she would be judged for promotion and tenure. "Excellence in" could describe the first category, and "highly competent" could describe the second.
• The categories should be realistically evaluated.
• A "mid-term" evaluation with the department chair is essential.
• The department head and colleagues should visit the classroom and the laboratory to authenticate the value of what the assistant professor is doing.
• There should be agreement that the amount of time spent reviewing grants and manuscripts should carry the same weight as the time devoted to creating materials to support classroom instruction.
• Professional societies should honor outstanding teachers.
• Articles extolling faculty who give the extra energy to teach

undergraduates should be published: Who are they, what motivates them, and how do they get their rewards?

• Every faculty member should be expected to teach, participate in scholarship, and serve.

• It should be recognized that support of excellence in teaching does not necessarily mean less time or effort on research and extension but a commitment to raising the quality of instruction and teaching programs to higher levels.

• A teaching foundation that parallels the research foundation found on most campuses should be set up.

• Research excellence is currently viewed and valued as bringing in money to run and enhance a department or college. Therefore, the value placed on research is in part due to a "greed" factor. If there were some way that research achievement could be divorced from indirect costs, it might be viewed in a more equitable light.

Strategies for the Management of Time

Vidaver and Kelman used their fourth point, strategies for management of time, as a summation of the various perceptions of conflict between teaching and research.

The discussion group agreed that as colleges of agriculture in land-grant colleges and universities continue to debate the balance of teaching and research as being imperative, we must ask ourselves, "Are the problems real or just a figment of our imagination, and are our perceptions pervasive throughout the university or just unique to the college of agriculture?"

In order to resolve the problem, we must do the following: develop clear expectations of faculty who teach, focus on synergistic activities that promote both teaching and research, and encourage classroom visitation to complement the student evaluation.

Define the responsibilities of the job. Have a formal commitment to all responsibilities without assigning labels; for example, require each faculty member to spend 10 percent of his or her work time in public service. Have and demonstrate an organizational philosophy that is clear and concise, one that supports good teaching and/ or good research and scholarship, and one that provides good mentoring for new faculty and effective leadership.

Rewarding Excellence in Teaching: An Administrative Challenge

William H. Mobley

Samuel H. Smith, Rapporteur

The session of the conference described in this chapter opened with the assumption that a balance between teaching and research is established as a clear need in the university. This chapter turns our attention to how to achieve that balance through administrative influence, action, and encouragement. Specifically, this chapter is concerned with the process of recognizing and rewarding the most effective teaching, thereby leading to the improvement of undergraduate programs in universities.

An organizing premise of the entire conference was that scholarship, research, and undergraduate education are interactive in the university. We have built our faculties on the basis of the fact that if one is improved, so are the others. Setting standards of expectation of anyone who teaches in the university and rewarding excellence in teaching does not mean a deemphasis of research and graduate education. It means reinstalling undergraduate education as a priority in the comprehensive university.

A challenge to presidents and provosts will be to ensure that deans, directors of experiment stations and extension services, and department heads recognize that undergraduates are vital to their self-interests. In a larger sense, the undergraduate program dictates the nature of the pipeline. It is the one component that must be concerned about future teachers, scientists, deans and directors, innovators and entrepreneurs, agriculturalists, and human capabilities.

We accept the principle that the university gains in intellectual strength through its research and scholarship. The reputation of the university grows as the extent of its research and graduate programs grows in quality and scope. That is the national focus. Society's support, however, particularly support by state legislatures, public support, and the support of economic institutions within the state, region, and country, is dependent more on the quality and success of the baccalaureate and professional program graduates. Undergraduate education represents the strongest constituent influence on legislators. Although doctoral graduates are a strong demonstration of the quality of a faculty and can help to facilitate aspects of the collegial and granting agency networks, relatively few doctoral graduates have helped us in the state legislature.

The challenge to the university administration lies in facilitating such a reemphasis of teaching. It is a matter of creating the appropriate climate or ambience. The administration must communicate its desires and intent to reward teaching. It will be only rhetoric until actions that include both extrinsic and intrinsic rewards confirm the administrative intent.

The university administration should recognize the fact that the faculty that will carry out teaching activities is already largely in place. The faculty is there, although faculty turnover—and perhaps growth—provides opportunities to set new directions. Most universities and colleges of agriculture have faculty members who want to teach. They take pleasure in their work and desire to do it well. It may be a matter of simply giving them the opportunity and time to do so. It is more likely, however, that it is also essential to recognize and reward those who want to teach and do it well and make them feel that their efforts are relevant and important in the modern university that aspires to preeminence. Too frequently we criticize poor teaching, but we fail to adequately recognize and reward good teaching.

The current system of evaluation and reward is based on the concept that the business of the comprehensive research university is developing and transmitting knowledge. (Increasingly, innovation is added to the mix of missions of the university as technology transfer becomes a greater goal of national policy.) The system of rewarding research is established by the scientific community. Peers reward research by evaluating and accepting publications for research journals. Grants are rewards for well-written ideas and peer-evaluated research. Promotion, tenure, and salary are rewards for those who have become competitive in the scientific and academic marketplace. Within the bounds of the university's goals and objectives, it is fairly easy to administer the rewards for research. It is more difficult to administer well the rewards for teaching.

The system for evaluating research is well developed and entrenched. Such evaluation is, in fact, a means by which universities compare themselves with each other. Exceptional teaching is

142

also recognized within all institutions. But universal standards for evaluating teaching similar to those used for evaluating research simply do not exist. It is more difficult to compare the teaching excellence in one university vis à vis that in other universities.

Teaching is not validated by a disciplinary "invisible" college of colleagues. A professor highly valued for his or her pedagogic abilities does not really have his or her value set by the academic market. Some sense of the knowledge of the recognized teacher may be obtained, of course, if the faculty member contributes to the synthesis of knowledge in books, book chapters, teaching modules, reviews, summaries, national or institutional study committees, and campus distinguished teaching awards. For the outstanding, well-loved professor, however, there is no universal prize that recognizes professional competence.

The reward system for excellence in teaching is, in fact, not a single issue. It involves a multiplicity of factors. The rewards for teaching must be tied into the academic programs, the entire reward system, the administrative philosophy, the physical environment, and the appreciation of the faculty's perception of the job. Understanding the last factor may be the most important.

Faculty members are each educated in depth in a subject matter or discipline. Few are educated in educational psychology, methods of instruction, or ways to change human behavior. They are employed because they have skills and special knowledge, and they probably will resist attempts to get them to perform in ways in which they are not equipped. At Texas A&M University, we have created the Center for Teaching Excellence to provide a focal point for enhancing faculty teaching skills, researching approaches that enrich the classroom experience, communicating effective ideas and approaches, and providing diagnostic, evaluative, and development services on the individual and departmental levels. One of our Capital Campaign goals is to endow this center, and we have found it to be a very attractive concept for donors.

Whatever we do, the lecture-seminar-laboratory-testing-grading method of instruction will continue to dominate university education. The single teacher—knowledgeable, experienced, and insightful—with a group of well-prepared students will continue to characterize the best of undergraduate and graduate teaching. Thus, the most important factor is the faculty member.

The rewards can be recognition, promotion, salary, and tenure. Also, they can be improvement of the environment of the teaching faculty: graduate assistants to help develop courses, word processors, assistance with slide preparation, and secretarial assistance. Or they can be titular recognition, pedagogical research support, travel to national meetings, sabbatical study, and time to focus on instructional development activities. We are well aware of the cynicism that can develop among our faculty and students when we give only lip service to excellence in teaching but reward only excellence in research.

The administrative challenge will also be to recognize the special problems of the faculty. Young tenure-track faculty members face immense problems. There is a bureaucracy with which they must cope. They must seek financial sponsorship. Many would like to teach, but they have uncounted pressures on their time. The system makes little allowance for cultural and personal problems. Minority faculty members face special problems brought on by both the general competitive aspects of faculty life and the need that they act as role models and mentors for the minority students and serve on a disproportionate number of search, academic, and other committees.

Although there is an extensive amount of literature critical of university teaching, particularly undergraduate teaching, the administration has had only scant guidance on what is truly feasible. The common recommendation for the evaluation of teaching is the periodic, regular student evaluation of the course. The greatest value of such an evaluation is probably that it has caused faculty members to think about their performances, course contents, methods of delivery, and cooperation with each other. The student evaluation may become less reliable—at least less approbative—when there is an intent to introduce greater rigor into the course. At present, however, there seems little that can replace it—nor should it necessarily be replaced—but it does need to be supplemented.

We have institutionalized the practice of student evaluation of teaching. Additional approaches, however, seem desirable. One might be reports of the faculty members themselves. Another might be peer evaluation of the ability of the faculty member to perform, for example, in a seminar, if not in a classroom. Another, too, might be the evaluation of materials prepared by the faculty member, or videotaping for self or peer critiquing of classroom presentations and interactions.

We are not looking for a radical change in the kind of teaching carried out in universities. What we are looking for is improvement in the effectiveness of the teaching—a reaffirmation of an old priority.

The primary factors that should be measured are those characteristics that we can expect faculty members to have:

• Rigorous academic preparation for the subject being taught. One technique to achieve this is for the faculty member to spend time in research and scholarship.

• The ability to prepare adequate test materials and assignments. Improvement of teaching lies in the demand for precise and demanding requirements of student performance; but experimentation, innovation, nontraditional courses, and encouragement of broader education are valuable as well.

144

• The characteristics of energy, ability to perform, enthusiasm, humility, generosity, honesty, and sense of reality and balance.

Add to these the encouragement, as far as possible, of the one-on-one relationship of each student with some mature, adult faculty member.

This is not a wide range of expectations, but they are the marks of teaching excellence that the university administration must strive to reward.

The answers will not come by climbing aboard some bandwagon. Until an accepted general formula is devised, the answers will likely come from innovative approaches that can be tested within each university. Several possible innovations may be worthy of exploration. For example:

• Differentiated career tracks for pedagogical professors and research professors. All would have teaching, research, and service expectations, but the mix would differ.
• A proportion of the faculty dedicated to undergraduate instruction as the primary dimension of the professional role.
• A separate faculty and reward system for servicing the core curriculum.
• Required course(s) and supervised teaching internships for graduate assistants.
• Dramatic reduction in the use of multiple-choice examinations and an increase in essay examinations and didactic instruction.
• Increased emphasis on matching learning styles and instructional methodology.
• Increased use of well-developed, individualized, interactive, programed instruction in basic courses.

I established the University Multiple Missions Task Force to examine such issues.

Administrators have an important role to play in helping to create expectations and reward systems that effectively reaffirm our commitment to undergraduate teaching in the multiple-mission, comprehensive research university.

RAPPORTEUR'S SUMMARY

A clear message was articulated by William H. Mobley during his comments concerning the rewarding of excellence in teaching. The message was that if an institution wishes to maintain or enhance quality in teaching, it needs to establish clear expectations coupled with a consistent reward system.

I have attempted to capture the tone of the discussion that followed the formal presentation and hope that the messages it contained are equally clear.

Teaching is a major function of any college or university. Unfortunately, over the past few decades, as research and public service have been included within institutional missions, the roles of teaching and teachers have become blurred. Many in our institutions were hired as teachers prior to this change and are now frustrated or angry. The rules of the game have changed. The perception is that rewards go to researchers or grant "getters," with little recognition of teachers' efforts.

The frustration that teachers feel in not being recognized for their teaching ability is altering the actions of newly hired or junior faculty members. Those new to academia quickly learn how to advance their careers. In many institutions, particularly the larger ones, the most rapid career path is often through research. It is not surprising, then, that younger faculty members shy away from teaching.

The problems relative to recognizing teaching are further complicated by the fact that institutional officials are constantly stating that they truly value teaching, but their reward system clearly shows that these statements are hypocritical at best. It has been suggested that the best means of valuing an institution's desire to reward teaching excellence is not to count the numbers of plaques given out annually for teaching performance but to compare the growth in budgets for teaching, research, and public service. During our discussion, a participant asked when was the last time anyone had heard a university president boast of recruiting an outstanding teacher? A participant also asked when we had last heard of an outstanding teacher being recruited away by another institution, being given a major salary increase, or being asked to bring his or her teaching assistants along when starting a new job?

Before we despair, however, the problem of rewarding excellence in teaching is relatively clear, and there are options. We looked first at how to evaluate teaching. The obvious way is that used for peer evaluation of research, which uses the numbers of refereed journal articles as a measure of quality or productivity. In our discussion, it became clear that peer evaluation of research is broader than the individual institutions and is closely linked with international professional standards. These superinstitutional standards were then contrasted to those for teaching, which tend to be linked most strongly to an individual institution. We need to recognize that research evaluation is largely globally based, and in most instances, teaching evaluations are generally local.

The local nature of teaching evaluations should be viewed as positive rather than negative. It is easier to affect local evaluation systems than it is to affect teacher evaluations on a global basis. We should not be anguishing over the lack of a widely accepted system of evaluations for teaching but, instead, attempting to improve the systems that exist in our own institutions.

The concept of valuing research by counting the numbers of research articles or grant money is probably more expedient than accurate. This means of valuing research, although probably not terribly accurate, is generally accepted. Discussion was particularly aimed at those who are searching for the perfect means of teacher evaluation, or the "Holy Grail," in the words of one of the discussion participants. It was strongly suggested that, although it is good to keep searching for the Holy Grail, we still need to develop an evaluation system that would be accepted at individual institutions. In other words, do not delay the establishment of an accepted evaluation system while looking for the Holy Grail.

An accepted means of teacher evaluation could be used to shape behaviors of faculty members as well as those of departmental, college, or university administrators. The institution would be communicating its expectations while it would be positively reinforcing behavioral patterns consistent with those expectations. During the discussion, the need to positively reinforce teachers was stressed, and the advancement of faculty members solely because of their research was lamented.

This discussion was, as a whole, realistic yet optimistic. It was recognized that the changing role of institutions of higher education has resulted in the creation of a role broader than teaching. This more inclusive role was not viewed negatively but was recognized as creating a need to set expectations in teaching and couple them with a consistent reward system.

18

Integrating Agriculture into Precollege Education:
Opportunities from Kindergarten to Grade 12

Harry O. Kunkel

Janis W. Lariviere, Rapporteur

Food, agriculture, and natural resources are factors that affect all people. The understanding of these factors, however, is not well integrated into general education in the United States. Their role in the economy, their vital function in the health and quality of life of people, and their importance in global interrelationships are scarcely recognized in general education. Moreover, the vitality of the progress of students from precollege to colleges of agriculture and life sciences is a key factor in the vitality of the professional missions of these colleges. It is important in discussing curriculum reform and revitalization that the precollege education also be reviewed.

In the interest of focusing the discussion, look at three aspects of the precollege education.

The first and most obvious aspect is the traditional integration of agriculture in precollege education, namely, the program in vocational agriculture. Companions to that are the 4-H Club and youth programs, which have a close association with the land-grant universities through the Cooperative Extension Service.

The second aspect, one of increasing importance, is the articulation of colleges of agriculture to students in science programs in secondary schools. This would provide a greater opportunity to bring so-called nontraditional students (for example, nonfarm, urban, and minority students) into agricultural education.

The third feature is the integration of agriculture into the elementary school curriculum. This may be the best means for very young people to relate food systems and natural resources to human life and well-being.

There are few studies that can give definitive answers to the persistent problems of trying to interest students in studies of agriculture in precollege and college settings. The information is more often anecdotal than empirical. One exception is the study of high school students' perceptions of colleges of agriculture and agricultural careers carried out by the American College Testing Program (ACT) (1989) for the Farm Foundation. That study reaffirmed suspicions that high school students have many misperceptions about agriculture-related careers and majors. The study also suggested that the student population interested in obtaining a major in agriculture in college is fluid: many students changed their objective away from a major in agriculture between the time they took the ACT test and the time they entered college. The implications are that education in and about agriculture can be translated into agriculture-related career goals at early ages, but competing interests have an impact as students make plans for their higher education.

Agricultural Education

A seminal study aimed both at vocational agriculture and agricultural literacy was a report by the National Research Council (NRC) (1988), *Understanding Agriculture: New Directions for Education.* That report found that agricultural education in secondary schools usually does not extend beyond the offering of a vocational agriculture program. Vocational agriculture has had a positive effect on thousands of people—students, families, and residents of the community—but until recently, enrollments were largely made up of white males; the program contents may be outdated and uneven in quality; and as a result of funding patterns, the education may be largely restricted to vocational education. The NRC report concluded that the focus of agricultural education should change, many revisions are needed, the quality of the program should be enhanced, high-technology instructional materials and media should be available, and all students should be enrolled in supervised experiences. The report also implied that agricultural education should contribute to general agricultural literacy.

Led by forward-thinking groups in vocational agriculture, state education agencies and individual schools across the nation are changing their programs in ways that reflect the recommendations of the NRC report. Flexibility is being introduced into the program. In Texas, courses are now offered in a semester-length format rather than as year-long courses set in rigid sequence. The program designation has been changed from "Vocational Agriculture" to "Agricultural Science and Technology." Supervised agricultural experiences are acceptable in a much wider range than the show animal or garden and are no longer called supervised "occupational" experiences. Any high school student can take a semester course as an elective, but only if the high school offers the program in agriculture. In Texas, honors courses that parallel first-year college-level offerings in animal science, agricultural economics, and agronomy and an experimental course on rangeland management are being offered. The inventory is 23 semester-length courses in such topics as food technology, agribusiness, management, wildlife, leadership, and agricultural production; three honors courses; and one experimental course.

In such a context, there is substantial transition. The emphasis is turning more to the education of students who are college bound or who seek employment other than farming. Vocational emphases are changing from skills with plants and livestock to skills of leadership, decision making, and technology. The content of the courses can become more rigorous and useful. Agricultural science and technology, stripped of its inflexible cloak, can open agricultural education to a much larger number of students. Enrollments in the agricultural science and agricultural business programs that were once vocational agriculture grew from 47,073 in the fall of 1986 to 64,681 in the spring of 1990.

The directive for the transformation at the national level is now the Strategic Plan for Agricultural Education, adopted by the several national interest groups and associations related to vocational agriculture education. The plan calls for updating instruction in and about agriculture, serving all people and groups, amplifying the "whole-person" concept to include leadership and interpersonal skills, fostering entrepreneurship, elevating standards of instruction, and developing educational programs that respond to changes in the market for students.

The essential key to driving these trends is the development of instructional materials. The closer that such development can be to the colleges of agriculture, food, and natural resources, the better the interaction will be. Someday, too, it may be possible for colleges of agriculture to couple with secondary school agricultural science by way of a satellite-mediated "distant education," with the home base being a land-grant university, for example.

Linkage With Science Teachers and Courses

Colleges of agriculture are also recognizing broader mandates. This is reflected in their names: college of agriculture and life sciences; college of agricultural and environmental sciences; college of agriculture, food, and natural resources; and so on. A number of colleges of agriculture are apparently making a strong move to capture the environmental bent. As that thrust continues, high school science, particularly biology, and teachers of science can become key factors in an articulation of the precollege education in agriculture and the natural resources. This connection becomes increasingly important as agricultural education in high schools and colleges reaches for diversity in the student "pipeline."

Science can be taught in the context of agriculture. For example, molecular genetics can be directed toward the conservation or improvement of food supplies. Agriculture and related subjects must be taught within the context of biology, chemistry, and the social sciences. Some colleges of agriculture have managed National Science Foundation-sponsored young scholar programs to demonstrate career opportunities in science. Some have provided specific courses, such as Ecology for Teachers. Some have provided teaching materials to teachers of biology and chemistry. Through interested teachers of biology—biology is a particularly appropriate subject because it is usually taught in the tenth grade—capable, interested students have been identified early and "adopted" by departments in the college of agriculture. The trend probably will be to integrate the sciences from kindergarten through grade 12, and colleges of agriculture, being multidisciplinary, should be in league with that spirit. As relevance, quality of education, and interest among teachers and students in secondary schools are directed toward agriculture and natural resources, word will spread and new linkages can be opened.

The Changing Student Pipeline

In the past three decades there have been unprecedented changes in the undergraduates who flow through colleges of agriculture, food, and natural resources. Student interests and attitudes have varied. Some patterns appear to be cyclical. The late 1960s and early 1970s saw large influxes of students with concerns about U.S. society. At the time, they were considered nontraditional students. We may now be seeing a return of large numbers of students with societal concerns that may override the preoccupation with jobs that was evident in the 1980s.

The students who are coming to colleges of agriculture through the science course route are different kinds of students and may have different philosophical bents. Many are interested in conservation issues, environmental law, and helping people. They will have practically no interest in operating a farm. Many appear to be looking for multidisciplinary outlets—which should fit colleges of agriculture—and they understand networking better than ever before. Many will come with the intent to obtain more than one degree, including the master of business administration, doctor of medicine, master of agriculture, and other professional degrees.

As the pool of prospective students changes, we are in a continuing search for words. The recruiting words now (again?) seem to be environment, conservation, science, job satisfaction, molecular genetics, health, and animals. Some students—as well as some of the faculty—who come to natural resource and other agriculture-related courses for traditional reasons may regard those students interested in conservation and the environment as espousing philosophies inconsistent with agricultural thought. But, it is a dichotomy that we should exploit, not fear.

All prospective students may now be beginning to focus more on gratification derived from technical work than on job pay. As a colleague in animal science, Gary Potter, notes, what is not "in" is trying to sell, justify, or defend an industry. Teenagers are not very interested in the reasons why it is important to grow food to feed the world. Many students take up agriculture-related majors in the university for reasons that make little professional sense: they like to ride horses, they like hunting and fishing, or some day they would like to operate a farm in addition to their primary employment. It is the environment outside the schoolrooms that may provide the greater influence toward interest in agricultural courses. Once in college, however, students can be taught both what they want to learn and what we would like them to learn.

Agricultural Literacy

The NRC report (1988) also recommended that, beginning in kindergarten and continuing through the twelfth grade, all students should receive some systematic instruction in agriculture. A persisting belief among those concerned about agriculture is that we not only teach skills and technology but that a body of education about agriculture should be provided to larger numbers of precollege students. This envisions value in general knowledge about the history and current economic, social, and environmental significance of the food and fiber system (National Research Council, 1988). Such understanding would include some knowledge of food and fiber processing, distribution, and domestic and international marketing. Such education might also include the practical knowledge needed to care for the environment—lawns, gardens, recreational areas, parks, and communities—that touches individual lives. Most importantly, the individual knowledge base should include enough knowledge of nutrition and food safety so that people can make informed personal choices about diet and health.

Historically, agricultural literacy for many has come through the national Future Farmers of America organization integrated with vocational agriculture education and the 4-H Club program, both of which have the basic infrastructures to support such activities. The

4-H Club program touches millions of students, and its value to the development of the individual young person is unquestioned. However, agricultural literacy, per se, seems to be a diminishing goal of the 4-H Club program, although many of the projects are related to biological phenomena and consumer skills, and they do contribute to students' understanding of the food and fiber system.

Programs have been devised that attempt to fill the void with the larger student population from kindergarten to grade 12. The Agriculture in the Classroom program of the U.S. Department of Agriculture, which is supported by state departments of agriculture and the farm bureaus, has developed some very good materials and reflects diverse efforts, particularly at the lower grade levels. The California-based Life Lab program, which incorporates agriculture into core subjects such as science, has demonstrated a possible mechanism of positive intervention in the elementary school curriculum. Projects Wild (for students from kindergarten through grade 12 sponsored by the Western Association of Fish and Wildlife Agencies and the Western Regional Environmental Education Council) and Learning Tree (for students from kindergarten through grade 6 sponsored by the Western Regional Environmental Education Council and the American Forestry Institute) are designed to enhance awareness of wildlife and forests in the school setting. Many teachers have plants and small animals in the classroom to aid the learning process. Children have a love of nature, which sets the stage for a closer connection to agriculture.

These programs also have their limitations, not the least of which is the fact that there can be too little incentive for elementary and secondary school teachers to use the materials. Teachers receive a lot of material; much of it is very good, but they have limited time to go through it all. Therefore, widespread use of agricultural literacy materials is not likely to occur unless these materials receive approval from state education agencies. Even though there is official acceptance, however, the focus must still be on teaching the essential elements, not agriculture.

Many symbols of agriculture that we have used in the past—the fluffy and cute (chicks, calves, or baby lambs), the grotesque (cockroaches), the beautiful (waving fields of grain, valleys of flowers, bucolic farm scenes)—to attract young people to the marvels of living things and the countryside may tell little of the real meaning of food and agriculture. It is imperative that we rethink the approaches. But tying a commodity, for example, corn, to the essential elements of education may work: teach addition and subtraction with tropism as the visual example, count corn kernels instead of other things, teach social sciences and history while asking why the Midwest became the Corn Belt or how certain cultures came to depend on corn as the staple food. At another juncture, one can teach students that everyone is dependent on the food system, that is, that the food system exists because it provides for the needs of people by maintaining health and well-being. One can teach history in the context of wars that have been fought over agriculture and the denials of food to people. One can teach students about microorganisms related to food safety and that bread, cheese, and pepperoni are fermented products.

Conclusion

We do not know whether any particular approach will work. However, education in and about agriculture from kindergarten through grade 12 is worth rethinking. At the outset I noted that it is important that people know that food, agriculture, and natural resources affect them. In order to gain acceptance of the principle, we may well need to put aside our parochial world views of agriculture, the desire to create a populace that thinks about agriculture as we want them to, and even the wish to rescue the traditional agricultural majors in college by turning around their decreasing enrollments. The growth of agricultural education in secondary schools, the movement toward integrating agricultural examples into the essential elements of education, and the enrollment of the diversity of students in colleges of agriculture may depend on colleges of agriculture also viewing themselves in truly flexible—and confident—ways.

References

American College Testing Program. 1989. High School Student Perceptions of Agricultural College Majors and Careers. Oak Brook, Ill.: Farm Foundation.

National Research Council. 1988. Understanding Agriculture: New Directions for Education. Washington, D.C.: National Academy Press.

RAPPORTEUR'S SUMMARY

Harry O. Kunkel began the discussion with an excellent overview of the problem. He stressed two goals for integrating agriculture into the precollege curriculum. People need to know about food, agriculture, and natural resources; and colleges of agriculture are dependent on the continued flow of precollege students into their programs.

The discussion continued with a series of insights from many participants about how these goals can be accomplished. As a high school teacher representative, I felt that colleges of agriculture must reach out to precollege teachers in a "hands-on" way if they really wished to have an impact. Mailings of information or even activities are rarely used unless the teacher has been to a workshop where they have done the activity. I also encouraged colleges of agriculture to have the teachers help to design the activities. I cited the Woodrow Wilson summer institutes at Princeton University as excellent models. They bring in secondary school teachers from around the country and present them with enough information so that they can design student investigations and disseminate them in their home states. Teachers have excellent networks for passing on information by presenting workshops at local, state, and national science teacher conventions. There were more than 10,000 science teachers at the most recent National Science Teachers Association convention in Houston, Texas.

Richard Reid, from the Society of American Foresters, mentioned two successful projects for teaching young students about natural resources that also directly involve teachers. These are Project Learning Tree and Project Wild. Project Learning Tree is for students from kindergarten through grade 6, while Project Wild has activities for students from kindergarten through grade 12.

Paul Williams, from the University of Wisconsin, encouraged the participants to bring their exciting agricultural discoveries into precollege classrooms. Williams has developed a strain of cabbages, Fast Plants, that allows students to investigate a plant's entire life cycle in one semester. He has set up a team partnership with teachers that encourages them to use these plants. These plants and activities with them are presented at teachers' conventions around the country.

Williams stressed that it is imperative to get agriculture into the curriculum of nonscience majors at the college level, because these people will be teaching the students in kindergarten through grade 6. Williams also encouraged colleges of agriculture to seek corporate support for teaching teachers. He has an ongoing project funded by the Kellogg Company that teams a science teacher with a vocational agriculture teacher for a 2-week immersion course on campus.

Dwayne Suter, of Texas A&M University, also advised session participants to seek corporate sponsorship. He mentioned the Consortium of Math and Science Teachers in Texas, which is partially funded by International Business Machines and other corporate sponsors. Suter encouraged the group's quest for ways to inform teachers about opportunities in agriculture by describing a study that showed that high school teachers are one of the three most important factors in a student's decision-making process regarding a career (parents and a person in the field being the other two).

J. Leising, of the University of California, Davis, felt that a cohesive framework of learning outcomes for each grade level would be helpful.

Recruitment of minority students is an important issue in this drive to interest more students in agriculture. Marquita Jones, of Indiana University of Pennsylvania, urged colleges of agriculture to have this as a focus. Urban and inner-city youth must be reached early. They need mentors at the precollege and college levels.

The next topic of discussion was the need for precollege teachers to do science themselves. Science can be deadly as a spectator activity. Paul Williams believes that there should be more teacher internships. G. Carlson, of Western Illinois University, stated that his institution has had many successful internships. He advised giving some college credit for the experience so that the teacher can work toward a higher degree.

The session continued with a number of helpful comments and suggestions. M. Hoppe, of South Dakota State University, praised an activity she had recently participated in—the Expanding Horizons Conference for Young Women. G. Sharma, of Alabama A&M University, suggested that colleges of agriculture need to help vocational agriculture teachers expand their horizons so that they can better prepare students for agriculture in the twenty-first century. S. Batie, of Virginia Polytechnic Institute and State University, asked whether there was a clearinghouse for information about all the excellent programs that were mentioned. Paul Williams suggested joining the National Science Teachers Association and the National Association of Biology Teachers. Their journals, *The Science Teacher* and *The American Biology Teacher*, are good sources for information about programs that are already available.

The group's discussion then turned to career education. High school students have a very limited knowledge of careers in general and even less of careers in agriculture. Colleges of agriculture need to get their message to these students. They can send speakers to science classrooms; bring teachers, counselors, and students to campus; or send materials with career information to biology teachers. I stated my opinion that although teachers would not do activities without hands-on experience, teachers would distribute career information. G. Carlson showed a brochure that Western Illinois University has produced to recruit people to agriculture. I suggested that brochures similar to this example should be given to biology teachers at high schools that would be likely sources of students for a particular university. A. Jones, of the University of Nebraska, stated that his institution had success with inviting teachers, counselors, and principals onto campus to visit the College of Agriculture. D. Hersey, of the University of Maryland, suggested that preservice courses for teachers should be an important target for agricultural career information and for information on agriculture in general.

B. Hooper, the director of the Association of American Veterinary Medical Colleges, reminded the participants that high school is too late for many students to hear about career opportunities. Many students have already denied themselves entry into a career because of their reading and math skills. Career education must start at a young age. Paul Williams suggested that agricultural science would fit very well into the new math literacy programs for middle-school students.

The session concluded with S. Maurice, of Clemson University, suggesting that we reduce the use of the term *agriculture* because of its negative image. I responded that many urban and suburban precollege students have no image of agriculture. If these are the areas from which colleges of agriculture hope to draw new students, the colleges need to decide what agriculture is today and bring that message to these students in a meaningful way.

Toward Integrative Thinking: A Teaching Challenge

Richard A. Herrett

Roy G. Arnold, Rapporteur

A recent National Research Council (NRC) committee noted in the study *Educating the Next Generation of Agricultural Scientists* (National Research Council, 1988) that there is an appalling lack of reliable data on educating agricultural scientists. They relied on the collective wisdom of the committee to fill that data gap. I have discussed the issue with several of my colleagues in industry, however, and have woven their thoughts into this discussion.

The NRC committee also noted several factors that influenced their thinking and that are equally important to this discussion:

• the rapid changes in several sciences critical to agriculture, such as human nutrition, forestry, and food and fiber processing;
• the future uncertainty of public- and private-sector investments in agricultural sciences and technological development;
• the economic adjustments, social and demographic changes, and institutional reforms; and
• the lack of information on how current educational policy and programs affect the quantitative dimensions of doctoral scientists.

These variables are still at work today and influence to varying degrees the considerations discussed here.

I approach this discussion from the vantage point that industry is the marketplace and that academic institutions provide a product to fit into that market. I do not mean to infer a process such as serving hamburgers at McDonald's, nor do I wish to suggest that I am minimizing the many other vital aspects of the university, such as research and extension. Industrial research and development is not done in a perfect world, and there are characteristics of industry that set it apart from the governmental and the academic worlds. These characteristics must be recognized.

I will examine some trends that are taking place and that should be considered as one defines the product for this marketplace. I will also examine some of the major changes taking place in our industry—agriculture—that influence the kind of product we need. Finally, I will examine what I perceive that these trends and changes tell us in the way of the challenges facing educational institutions today.

An examination of some of the circumstances that have influenced my thinking may be helpful as I attempt to paint my views. My degrees, both undergraduate (Rutgers) and graduate (University of Minnesota), were obtained from more or less traditional land-grant universities.

In both universities there was a college of agriculture and a main campus. The latter was, in the estimation of some, where the real education took place, and there was a major delineation between the two facilities. Indeed, even within each of the institutions there were incredible walls and a sense of isolation between various departments. There was very little evidence of any exchange of ideas or people.

My entire professional career has been spent in the industrial world, even during an initial period at the Boyce Thompson Institute (BTI), where I was a Union Carbide industrial research scientist working in an academic environment. Although I had the good fortune of studying under professors such as E. C. Stakman, P. D. Boyer, J. J. Christensen, and others, it was at BTI when the international dimensions of plant biochemistry and the opportunity to participate in seminars, workshops, and symposia took on an entirely new dimension in which it was possible to integrate one's own training with other disciplines and to begin to sense the powerful potential of modern science. This was especially so in the post-*Sputnik* era, which witnessed a tremendous expansion of scientific endeavor with relatively easy access to funding. It was indeed a glorious time to be a new doctoral scientist.

It was too good to last, of course, and it was there that I had my first exposure to the notion of flexibility. Union Carbide decided to relocate from BTI "to focus their research," and from that point onward "industrial research" became an identifiable activity. I should note that there was a strong academic bias against industrial research as part of a protective, supportive mechanism to retain the best graduates within the university system. Undoubtedly, that bias was related partly to the unusual demand for scientists at that time and partly to the nature of industrial research.

159

Characteristics of Industrial Research

The following are some of the obvious characteristics of research in a very generalized industrial setting. Indeed, each industry and each company within each industry will operate differently, which will affect these characterizations.

- short-term profit horizons, usually within 10 years, but sometimes within one or two quarters;
- little energy directed toward new knowledge, which is usually accidental if it is discovered and is generally difficult to develop;
  - susceptibility to economic dislocations;
  - continuous process of refocusing and restating objectives;
  - small or narrow discipline base; and
  - integration of several disciplines.

Trends in Industry

In addition to the more or less generalized characteristics of industrial research, there are certain trends taking place in industry that also influence the nature of the marketplace:

- consolidation—for example, in the past 20 years, the agrochemical business has gone from 36 to less than 12 major research and development-driven companies;
- new start-ups—entrepreneurial enterprises are often based on a new discovery or technology;
- integration of several disciplines—for example, biotechnology, which is multidisciplinary;
- new dimensions—for example, biotechnology, which includes ethics, strategic planning, economics, and communications;
  - increasing costs and regulations;
  - increased environmental awareness (costs); and
- expanded multinational dimensions—personnel, markets, and regulation.

These trends also have an impact on the marketplace and the type of product emerging from the university system.

Growing Environmental Awareness

Another example of the changes taking place is the growing awareness of the environment and the implications of this awareness not only on agriculture but also on the general public. This awareness clearly must be an integral part of the product emerging from the system. Increased environmental awareness is seen in education and politics as well as production agriculture. Whereas schools were formerly called colleges of agriculture, today they are known as colleges of agriculture and natural resources, among other names. The 1990 Farm Bill shows environmental awareness at an increased level over that in the 1985 Farm Bill, and production agriculture has gone from being chemically intensive to being knowledge intensive. For production agriculture, the change has been evolutionary and not revolutionary. The latter implies chaos. In the political arena there was concern that agriculture was part of the problem and needed to be controlled. Unresolved issues such as groundwater contamination, food safety, and sustainability all address those concerns. There can be little doubt, however, that farmers are ardent environmentalists. Farmers recognize the essentiality of sustainability as being integral to their economic survival and will therefore fight to extremes to preserve their land or enhance it for their children. The question becomes one of public perception fueled by the absence of understanding of what agriculture really is all about.

Challenges to Academic Institutions

These trends in industry and changes in environmental awareness and perception present challenges to the academic institutions. As I see it, these challenges can be drawn up into three categories: science, thought process, and communications.

Science

There are increasing demands for ever more specific knowledge and more sophisticated techniques. This is not surprising given the explosive growth in scientific knowledge and increased complexities in instrumentation. Science has become big business, which seems to place increasing demands on discipline-based training. Ironically, perhaps, there is also a need to focus on the big picture. This is brought about by demands that include the cultural, ethical, and economic implications of the impact of the science. Society no longer merely accepts the notion that all new technology is good and therefore good for people. Scientists must be trained to integrate their discoveries into the big picture. This clearly implies that there must be an integration of information from a variety of sources, and one trained in a disciplinary style simply cannot respond to this challenge. Indeed, the focused disciplinary approach does not produce a product that is well equipped for the industrial market.

Thought Process

There is little question that we can train people to assimilate factual knowledge simply by repeating it over and over again. The challenge, however, is to provide people with the ability and the skills to integrate and synthesize disparate pieces of knowledge that lead to the formulation of new conclusions from that knowledge. It is no longer adequate to learn the methodology of mapping genes and establish a picture of the human genome by using three-letter base pairs without an understanding and awareness of humans, human evolution, and human society. We simply cannot reduce the process to one of numbers. This challenges the educational system to go beyond sheer memorization to become skillful in the thought process.

Communications

There is an increasing knowledge base as the amount of information grows. This challenge mandates an ability to simplify and communicate complex ideas to an untrained but, by and large, intelligent public.

Conclusion

Although there are no hard numbers to cite, there seems little question that our industry—agriculture—is undergoing massive changes, changes that affect every aspect of that industry and, perhaps most significantly, the educational component. These changes place a premium on flexibility, or the ability to adapt to change. This is certainly true within industry, which has witnessed unprecedented consolidation and restructuring, especially over the past decade. The advent of biotechnology may be one of the most significant events of recent times, not only because of the new opportunities it creates but because of the potential impact on training, because it demands an integration of disciplines and skills that transcends the past tendencies to become compartmentalized. The changes in the constituent relationship from an industry based on chemical inputs to a knowledge-intensive industry—one that is perceived as the solution to the nation's environmental concerns—will place extensive demands on communications skills. The future success of America's number one industry—agriculture—will depend on the extent to which the academic community is successful in meeting these challenges.

When Theodore L. Hullar, chairman of the Board on Agriculture, considered the current problems facing agriculture, he highlighted the challenges of issues such as national and global public education about agriculture and the appropriate training of the next generation of international agricultural scientists. I submit that our very survival as the world's leading producer of high-quality food and fiber in an increasingly competitive, rapidly changing international market depends on our ability to meet this teaching challenge.

Reference

National Research Council. 1988. Educating the Next Generation of Agricultural Scientists. Washington, D.C.: National Academy Press.

RAPPORTEUR'S SUMMARY

The discussion following the presentation by Richard A. Herrett focused on some successful teaching approaches to the development of integrative thinking. Challenges and constraints were also identified.

Participants shared the following as being successful teaching approaches for the development of students' integrative thinking skills:

• freshman-level issues course in which students work with several different paradigms;
• introductory general education courses;
• small discussion groups involving students with widely disparate interests and perspectives;
• the broadest possible diversity in courses by including students with various backgrounds and majors;
• coupling of cross-disciplinary and cross-cultural courses that students take in pairs, with opportunities for group discussion of contrasting perspectives;
• capstone courses in each major;
• senior core course for students of all majors, focusing on the views of "adopted thinkers" on specific topics or issues;
• industry visitations and executive in the classroom-type programs for real-world exposure;
• internships, with advance preparation, monitoring, and follow-up reporting and discussion;
• mathematical models, which are integrative in and of themselves;
• decision-case approaches, by which the teacher lists a set of assumptions and calls upon the students to respond to those assumptions;
• goal-setting exercises;
• writing assignments;
• emphasis on listening skills (active listening);

• integration of subject matter around a common focal point, such as the environment and food safety; and

• careful development and analysis of course objectives requiring thinking skills, including consideration of appropriate final examination questions to assess the attainment of each objective.

Several challenges and constraints to the development of integrative thinking skills among students were identified. These included the following ideas and statements from participants in the discussion:

• problems in understanding what we mean by integrative thinking; Herrett's observations regarding different approaches to thinking in universities and industry are indicative of the problems of common definition and understanding;

• most of us are from "cells" (i.e., disciplines) that are becoming more specialized;

• disciplines were created by us, and we're comfortable with them;

• we are unable to achieve integrative thinking in students unless we display it in the classroom and are rewarded for doing so;

• we tend to focus on science, while many of our students are more oriented to agribusiness;

• most college faculty have had no formal preparation for teaching, but they are expected to be innovative; we have not been taught nor have we observed effective approaches to teaching integrative thinking skills; and

• large student numbers present a particular challenge for many of the integrative thinking approaches to teaching.

From the discussion in this group, it is clear that a greater investment in faculty development for teaching is needed. Faculty need to have opportunities to learn new concepts, develop new techniques and skills, and think about and plan their teaching approaches. Development of integrative thinking skills in students will require both faculty initiative and investment in faculty development.

The challenge is how to prepare students for a 40-year or more occupation involving seven or eight career changes. Many of our graduates' future jobs have not been defined at this time. Our focus should be on preparing students for life, not a first job. Integrative thinking should be part of that preparation.

Finally, it is important to note that we have not produced large numbers of failures. Most graduates of colleges of agriculture are quite successful, flexible, and adaptable. They seem to keep learning and growing throughout their careers. We can do better, of course, but let us not forget that we are not doing too badly at present.

Striving Toward Cultural Diversity

Edward M. Wilson

Peggy S. Meszaros, Rapporteur

Central to the future successes of our colleges of agriculture is how we adjust to the changing ethnic and racial composition of our student body. As the provocateur, I will introduce several issues in order to stimulate discussion, in light of the changing ethnic and demographic picture. I will suggest that our college curriculum emphasizes critical thinking instead of a focus on technical knowledge and touch on the need for foreign language competency, cognitive diversity, development of a global perspective, ethnic studies courses and programs, and an effective socialization process.

Changing Demographics

Demographic changes, ethnic diversity, global competition, and fundamental changes in the U.S. economy are a few reasons why we must reexamine the focus of our agricultural curriculum, educational philosophy, and the total academic environment. In *Workforce 2000*, Johnston and Packer (1987) project that, by the year 2000, nonwhites will make up 29 percent of the new entrants into the labor force. Immigrants will represent the largest share of the population increase, and almost two-thirds of the new entrants into the work force between now and the year 2000 will be women. Nonwhites, women, and immigrants will make up more than five-sixths of the net additions to the work force between 1985 and 2000.

Data from the 1990 census support this prediction and reveal that over the past decade the number of Asian Americans increased by nearly 108 percent, the Hispanic population grew by more than 50 percent, and those identifying themselves as American Indians grew about 38 percent, while the number of black Americans increased by 0.4 percent. These data not only support the *Workforce 2000* projections but also underscore the rapid pace of the nation's changing ethnic demographic picture.

A U.S. Department of Education report (National Center for Education Status Survey Reports, 1990) released in June 1990 correlates well with the 1990 census data and with the *Workforce 2000* projections. It shows a substantial minority pool among currently enrolled students. Over the 11-year period from 1978 through 1988, the number of minority students enrolled in college increased by 97.9 percent. The fastest-growing group was Asians and Pacific Islanders, which increased by 111.5 percent, followed by Hispanics at 63.1 percent and American Indians and Alaska natives at 19.2 percent. Blacks showed the slowest growth, increasing by 7.2 percent. This minority enrollment trend is particularly striking when one looks at the fall 1990 freshman class.

At a number of universities, freshman students of all minorities now form the majority. For example, Denise K. Magner reported in the November 14, 1990, issue of the *Chronicle of Higher Education* that from 1980 through 1989, the white undergraduate population at the University of California at Berkeley shifted from 66 to 45 percent. In the current freshman class, 34 percent of the students are white, 30 percent are Asian Americans, 22 percent are Hispanic, and 7 percent are black. Berkeley may not be typical, but it provides a barometer of the trend.

The ethnic makeup of the student body must be a primary consideration as we review and revise our undergraduate curriculum, contemplate the need for U.S. agriculture to become more competitive, examine the global nature of our economy, and strive to develop a harmonized multiethnic, multicultural campus environment. These enrollment shifts provide both opportunities and challenges.

Critical Thinking Versus Technical Knowledge

With the evolving multiethnic student enrollment, along with the need for U.S. agriculture to become more competitive in a global economy, curriculum revitalization should emphasize as its main purpose teaching critical thinking (i.e., how to think). This is especially important to a large number of minorities who are often underprepared for college. What we should pursue is an educational process that will teach our students to think and reason, which will improve their employment and earning prospects, add to their poise and deportment, develop their judgment, and round them out for a fully successful and happy life.

It is virtually impossible to make technical curriculum changes to keep up with the estimated 40 percent annual increase in technology and scientific knowledge. However, one suggestion for responsiveness is that curriculum review and revision should concentrate less on technical knowledge and more on the liberal arts, with a focus on communication, problem solving, the international and interdisciplinary perspectives of agriculture, foreign languages, and concepts that enhance lifelong learning. When they are carefully honed, these skills will provide students with the capacity to better cope with the knowledge and technology explosion.

Foreign Language Requirement

Foreign language competency becomes increasingly important as we increase our participation in the global economy. Evidence suggests that competence in a foreign language is more easily attained if it is introduced before the college level. Colleges of agriculture should consider requiring 2 or 3 years of high school foreign language credits for admission and requiring all agricultural majors to gain competence in at least one foreign language.

Cognitive Diversity

The increasing ethnic diversity not only supports the need for us to examine what we teach but also, and equally importantly, who and how we teach. We must critically analyze the teaching and learning process, which may provide insight and remedies for reducing the high dropout rates among members of some minority groups. We must recognize that not every student learns in the same way and that different teaching approaches may be necessary. We are reluctant to examine the concept of cognitive diversity and to accept the fact that we may have students in our classes whose cognitive styles reflect their varied cultural backgrounds.

James Anderson, professor of psychology at Indiana University of Pennsylvania, points out that cognitive styles are greatly influenced by the cultural values of individual ethnic groups and by socialization practices (Anderson, 1987). For example, Euro-Americans are influenced more by Western values and culture, while black Americans are influenced more by non-Western values and culture. The differences between these two cultures have created difficulties for some black students in our major universities, not because of a lack of ability but because they have not learned the skills and habits intrinsic in the Western culture. Anderson suggests that secondary and postsecondary school curricula must create a new awareness of multicultural, multiethnic characteristics, especially those that enhance the learning environments. He points out that minority students are successful in science programs that develop cognitive and noncognitive profiles of students, have preentry

167

programs, encourage bonding between faculty and staff, foster ethnic identity, and maintain high standards of excellence.

Annette Kolodny, dean of the faculty of humanities at the University of Arizona, points out that in any population, however homogeneous, one can always find evidence of different intellectual talents and cognitive patterning (Kolodny, 1991). But it is also true that different cultural groups may emphasize one cognitive style over another, for example, reasoning by analogy instead of a strict linear logic, problem solving through an inductive rather than a deductive approach, and learning through an empathetic identification with people rather than through abstract principles.

One may also say that science methodology is not the absolute teaching method for colleges of agriculture and that the laws of logic are not the laws of understanding. We can teach by taking into consideration the differences in the ways that people see the world.

Kolodny suggests that new instructional technologies allow faculty members to design programs, software, and classroom strategies tailored to a full spectrum of cognitive styles. In designing new teaching strategies, faculty members can experiment with technologies that appeal to cognitive styles other than their own.

Global Perspective

Another aspect of cultural diversity is the presence of increasing numbers of foreign students on our campuses. This provides a unique opportunity for our universities to add a global perspective to the learning environment. Foreign students can provide firsthand knowledge of a country's language, culture, history, economy, geography, and politics through formal and informal settings with students and faculty. These opportunities abound on most campuses as the number of foreign student enrollment continues to increase. In the academic year 1989–1990, the United States hosted 386,000 foreign students, a 5.6 percent increase over the previous academic year.

Colleges of agriculture should also consider a study abroad experience as an option for expanding the multicultural awareness of undergraduate students.

Ethnic Studies

The presence on our university campuses of significant numbers of ethnic minorities, Asian Americans, black Americans, Hispanic Americans, and American Indians, does not by itself constitute a multicultural, multiethnic education system. The university must take positive steps to ensure the respect for and sharing of the cultural values of the various ethnic groups and for the socialization of all students. If they are left without guidance or direction, students will gravitate to their own demographic group and look at others in terms of ethnic identity rather than individual characteristics. This quickly undermines a sense of community on campus. The practice of establishing singular ethnic studies courses and programs tends to segregate rather than unify the student body. The cultural values of different ethnic groups should be woven into the fabric of the component elements of the curriculum and not isolated in special courses.

The Socialization Process

Universities have the opportunity and the obligation to design a socialization and educational process that draws ideas and contributions from all sections of our heterogeneous society into a unifying core curriculum. The process of education and socialization should give recognition to our differences and individuality; however, emphasis should be on the common ground, that is, on the ideas, values, and norms we share as Americans. The undergraduate agricultural curriculum must prepare our students for the changing environment of U.S. and international agriculture. A base of technical knowledge is necessary; but it should be bolstered by communications, problem-solving, and foreign language skills as well as multicultural awareness. Students should join the work force with learning skills that enable them to benefit from the annual explosion of scientific knowledge.

Summary and Conclusion

The demographic shifts require that faculty members be sensitized not only to curriculum review and revision but to who and how they will be teaching. The university should conduct workshops to educate faculty and staff members about the history, culture, and cognitive styles of all ethnic groups. The faculty should develop a wide range of teaching styles to support the varied cognitive styles. Programs should be designed to guide students through a socialization process and help them appreciate the values of other cultures and reduce the culture shock that many students feel when they come to a campus.

The changing student demographics in U.S. universities provide the richest cultural diversity in the world. This diversity should be used to enhance the educational process; to create a sensitivity, awareness, and appreciation of different cultural practices and values; and to develop a sense of a multicultural community. No group should abandon its own culture; instead, the various groups should use their differences to contribute to the strength and unity of the multiethnic, multicultural community, not to divide it.

References

Anderson, J. 1987. Enhancing the research skills of minority students through knowledge of their culture and cognitive assets. Address delivered at the Seventh Biennial 1890 Research Symposium.

Johnston, W. B., and A. H. Packer. 1987. Workforce 2000: Work and Workers for the 21st Century. Indianapolis, Ind.: Hudson Institute.

Kolodny, A. 1991. Colleges must recognize students' cognitive styles and cultural backgrounds. Chronicle of Higher Education, February 6, 1991, p. A44.

Magner, D. K. 1990. Amid the diversity, racial isolation remains at Berkeley. Chronicle of Higher Education, November 14, 1990, p. A37.

National Center for Education Status Survey Reports. 1990. Trends in Racial/Ethnic Enrollment in Higher Education: Fall 1978 through Fall 1988. Washington, D.C.: Office of Educational Research and Improvement, U.S. Department of Education.

RAPPORTEUR'S SUMMARY

Edward M. Wilson, the session provocateur, began the session by asking how colleges of agriculture and related disciplines will adjust to the changing ethnic content of the student body?

Demographic changes force a reexamination of today's curriculum. Wilson outlined five major areas for reexamination: (1) critical thinking, (2) foreign language, (3) global perspective, (4) cognitive diversity, and (5) ethnic study programs.

The demographic changes driving our reexamination were highlighted in two recent reports, *Workforce 2000* (Johnston and Packer, 1987) and *Trends in Racial/Ethnic Enrollment in Higher Education* (National Center for Education Status Survey Reports, 1990). Both reports emphasize the increase in nonwhites, women, and immigrants as new entrants to the work force. By the year 2000, the U.S. work force will look very different from that of today. Minority students are now the majority at many universities. These enrollment shifts provide challenges and opportunities.

Wilson challenged the group to consider examining their college curriculum and asking whether critical thinking is emphasized. The importance of this skill for minority students outweighs the emphasis on technical skills. He further challenged the group to have colleges require 2 or 3 years of a foreign language. We must prepare students to understand different cultures, and understanding another culture's language is a step toward these goals.

College curricula must be examined for their emphasis on global perspectives. Foreign student enrollments are increasing, and both domestic and foreign students must realize that they live in a global economy. Study abroad experiences should be an expectation for all students.

Cognitive diversity is a curriculum dimension of great importance today. All faculty must analyze their teaching styles and student's learning styles. Cognitive styles reflect cultural values and the socialization of students. Science methodology is not the only important approach. Tools are available to faculty for diversifying their teaching methods.

Wilson was adamant that universities must take positive steps toward developing a multicultural perspective in students. He believes that singular ethnic study courses segregate students and that the better approach is to weave ethnicity into one approach, with emphasis on the common ground. This is crucial for undergraduate students.

The group discussion focused on several issues of great concern on campuses nationwide:

- Recruitment and retention of minority students are needed.
- The perception of agriculture in the minds of minority students is a major problem.
- University faculty must be sensitized to the demographic shifts. Many are not living in the twenty-first century.
- More minority faculty must be recruited, because role models are crucial to recruiting and retaining minority students.
- Corporate support is needed for scholarships and internships.
- Colleges have culturally biased curricula. Everyone should examine course content to be sure that we know what is being taught.
- Land-grant universities have a rich pool of foreign students. We should use them to provide experiences for all students. One participant gave the example of a dormitory for living and learning in which both domestic and foreign students are roommates.
- An example of a student organization, the Society for Minorities in Agriculture, Natural Resources and Related Sciences, has been formed. Its goals are to put more students in the pipelines and to help departments understand who they are teaching.
- A participant gave the example of an agronomy club sponsoring a multicultural event for which foreign students prepared their native foods; this is a way of exposing students to a multicultural perspective.
- A representative of the U.S. Forest Service gave an example of recruitment through special scholarships and internships, which has great promise for attracting women and minorities.

In the end, session participants realized that we are striving toward cultural diversity but that we have a long way to go to achieve it.

171

References

Johnston, W. B., and A. H. Packer. 1987. Workforce 2000: Work and Workers for the 21st Century. Indianapolis, Ind.: Hudson Institute.

National Center for Education Status Survey Reports. 1990. Trends in Racial/Ethnic Enrollment in Higher Education: Fall 1978 through Fall 1988. Washington, D.C.: Office of Educational Research and Improvement, U.S. Department of Education.

21

Designing an Environmentally Responsible Undergraduate Curriculum

Robert J. Matthews

Richard H. Merritt, Rapporteur

Farmers have historically occupied a place of pride in U.S. culture. Thomas Jefferson described farmers as "most valuable citizens . . . the most vigorous, the most independent, the most virtuous" (Peters, 1975:384). Jefferson located the source of these virtues in the special relation of farmers to their land: land ownership, he believed, gave them a vested long-term interest in their land, their communities, and the country. Given the immobility of land, their major capital investment, farmers could hardly afford to be anything but good stewards of their land, good citizens of their communities and the country. Agrarian writers throughout the span of American history have repeatedly praised farmers' stewardship of the land: Farmers might violently wrest their fields from the forest and prairie, the bounty of their crops from the furrowed soil; however, they lovingly protect the land from which this bounty springs.

Even the briefest survey of our nation's history reveals that the reality does not accord with the agrarian myth of good stewardship. Whether it is a matter of the eroded badlands of Oklahoma, the Dust Bowl era immortalized in John Steinbeck's *The Grapes of Wrath*, the massive destruction of hundreds of thousands of acres of wetlands in the San Joaquin Valley of California, the depletion of the Ogallala aquifer of the High Plains, the fertilizer-induced eutrophication of the Chesapeake Bay, or the salination and siltation of the Colorado River in Arizona, the conclusion is inescapable: U.S. farmers have not been good stewards of the land. U.S. farming practices, like many farming practices elsewhere in the world, have been and continue to be environmentally quite destructive. The most notable of these destructive effects have been (1) environmental contamination (pesticides, fertilizers, siltation, salination, animal waste, etc.), (2) habitat destruction (e.g., draining of wetlands, and deforestation), and (3) resource depletion (soil erosion, water depletion, etc.).

Water pollution is the most damaging and widespread effect of agricultural production. Agriculture is the largest nonpoint source of water pollution, which accounts for about half of all water pollution. Precipitation and irrigation-induced runoff carry sediments, minerals, nutrients, and pesticides into rivers, streams, lakes, and estuaries. Sediment deposition and nutrient loading are the most serious of these polluting effects.

Soil erosion and the subsequent sediment deposition in surface waters have been serious environmental problems for many years. Sediment deposition fills reservoirs, clogs waterways, and increases the costs of water treatment. Agricultural production is estimated to account for at least 50 percent of the sediment deposited in streams, rivers, and estuaries. The continuous monocultural or short-rotation farming methods currently used by U.S. agriculture is known to increase soil erosion; however, the increasing use of conservation tillage practices (which leaves crop residues on soil surface as a mulch cover) over the past decade has served to mitigate these damaging effects. Nutrient loading, however, remains a major problem, one that has been exacerbated in recent decades by the heavy use of synthetic nutrients. Nutrient loading stimulates algal growth in surface waters, which, when the algae die, depletes the available oxygen in the water and thus reduces (through death) aquatic plant and animal populations. It is estimated that 50 to 70 percent of all nutrients that reach surface waters have their origin in fertilizer or animal agricultural waste.

Pesticide contamination of surface waters has become a serious environmental and health problem over the past four decades. Between 450 million and 500 million pounds (204 million to 227 million kilograms) of pesticides, mostly herbicides, are applied to row crops each year in the United States. A small but nonetheless significant percentage of these pesticides is carried by runoff to surface waters, where they both harm wildlife and contaminate public water supplies. The bioaccumulation of DDT (dichlorodiphenyltrichloroethane) in fish, for example, contributed to the dramatic decline in predatory bird populations, such as the peregrine falcon, osprey, and bald eagle, during the 1960s and 1970s. Many of the public water supplies drawn from surface waters in states with significant agricultural production, for example, Iowa and Ohio, have detectable levels of pesticides. (One study [National Research Council, 1989] found detectable levels of two or more pesticides in 82 percent of the supplies tested; it found detectable levels of four or more pesticides in 58 percent of the supplies tested.)

Groundwater contamination by agricultural chemicals has also become a serious problem. Pesticides have been detected in the groundwater of 26 states. The most commonly detected compounds have been the insecticide aldicarb, which is the most acutely toxic pesticide registered by the U.S. Environmental Protection Agency (EPA), and the herbicide atrazine, which is oncogenic in laboratory rats and is currently under EPA review. The third most commonly detected pesticide is the herbicide alachlor, which is banned in Canada and is now classified by the EPA as a probable human carcinogen.

Fertilizer-induced contamination of groundwater is also a serious problem. The U.S. Geological Survey conducted a survey of 1,663 counties in the United States and found some 474 counties in which 25 percent of the wells tested showed elevated levels of nitrate-nitrogen attributable to nitrogen fertilizer use. In 87 of these counties, at least 25 percent of the wells had nitrate-nitrogen levels that exceeded the EPA's interim standard for nitrate in drinking water. The U.S. Department of Agriculture has calculated that some 46 percent of all U.S. counties contain groundwater susceptible to contamination from agricultural pesticides or fertilizers.

Current irrigation practices seem no less threatening. Irrigation has made agricultural production possible in many areas of the United States where it would otherwise be impossible, notably in the arid Southwest; it has also made intensive crop production possible in areas that could not otherwise sustain such production. But these production gains have been costly, not simply to farmers but to society as a whole. Irrigation has contributed significantly to aquifer depletion in many parts of the Midwest and West. In the arid West, where irrigation inevitably leads to salinization and mineralization of the soil, groundwater depletion has been accompanied by significant surface water and groundwater contamination, not to mention damage to the land under irrigation. These costs seem all the more unsupportable in the face of growing municipal and industrial demand for water coupled with the fact that much irrigated acreage is currently being used to produce surplus commodities (e.g., cotton, corn, and sorghum).

The agrarian myth of good stewardship makes it very difficult to formulate, much less implement, an effective public policy to deal with these destructive environmental impacts. Public policy develops only in response to a perceived need for policy, but the agrarian myth of good stewardship blinds us to the need for such a policy. It is hardly surprising that production agriculture is commonly exempted from statutes that might otherwise serve to ameliorate environmentally destructive agricultural practices. The agrarian myth of good stewardship makes it equally difficult to design, much less implement, a curriculum that might help to ameliorate these destructive environmental impacts. There is understandably little enthusiasm for curriculum changes that do not address a recognized problem.

The fact that a session entitled "Designing an Environmentally Sensitive Curriculum" was included among the conference's concurrent discussion sessions is a hopeful sign. It may finally signal both an end to the agricultural establishment's long resistance to acknowledging agriculture's environmental problems and a recognition of the need for colleges of agriculture to modify the undergraduate curriculum of their students in ways that might prepare them to address these problems. For too many years, colleges of agriculture have regarded their environmental science curriculum as little more than either an exercise in public relations "damage control" or a panacea for sagging undergraduate enrollments.

Having provided a reminder of what many already know, namely, that U.S. agriculture, for all its virtues, has real environmental problems, I would like to sketch out a broad curriculum that might better prepare our graduates to deal with these problems. My aim here is not to actually present the design of such a curriculum but simply to provide a framework for our subsequent discussion. The questions that I propose to address are these: (1) What should be the *goals* of an environmentally sensitive curriculum? (2) What are the *means* by which these curriculum goals should be achieved? (3) What *support* is necessary to ensure the success of a curriculum that undertakes to achieve these goals by these means?

Goals: Understanding and Analytical Skills

The title of the session at which this chapter was presented envisions an environmentally sensitive curriculum, as if all that students need is a bit of sensitivity training. Agricultural professionals need more than environmental sensitivity; they, like all other Americans, need to become environmentally responsible. An environmentally responsible curriculum, I argue, should provide two things: relevant understanding and appropriate analytical skills. Such a curriculum should provide students with an understanding of both the scope and the magnitude of agriculture-related environmental problems and of federal (and perhaps state) regulatory policies that are intended to mitigate these problems. Such understanding is essential if students are to develop the requisite sensitivity to environmental issues; however, it can hardly be the sole goal of a curriculum that aims to prepare students for the future. Problems and policies change. Students must be provided with the analytical skills necessary to (1) understand such new environmental problems as may arise and (2) understand, evaluate, and indeed, help to shape the new policies that will have to be devised to deal with these problems. Let me explain what I mean by "understanding" and "analytical skills."

176

By an understanding of agriculture-related environmental problems and policies, I do not mean simply a cataloging of current problems and policies. Students must, of course, learn certain facts, for example, that U.S. agriculture is heavily dependent on chemical inputs (fertilizers, herbicides, insecticides, fungicides, antibiotics), that pesticide contamination of surface water and groundwater supplies is quite widespread, that many of these pesticides are known carcinogens, that agricultural pesticides are regulated under the Federal Insecticide, Fungicide, and Rodenticide Act, that agricultural use accounts for 38 percent of water consumption in the United States, that the health risk from pesticide residues on foods is probably insignificant compared with the risk from naturally occurring alpha-toxins, and so on. But to have such knowledge is to fall far short of the sort of systematic, that is, holistic, understanding that I have in mind. If students are to appreciate not simply the magnitude and scope of agriculture-related environmental problems but also the practical difficulty of getting a handle on these problems, they must understand the historical, social, political, and economic contexts within which they have arisen. They must, for example, understand the changes in the farm economy and farm structure that have both occasioned and resulted from modern agriculture's dependence on chemical and other purchased inputs. They must also understand the impetus for these changes provided both by federal farm policies and by the research and education provided by publicly funded institutions, notably, the nation's land-grant universities.

The analytical skills that students need to acquire are (1) *problem recognition skills*, that is, the ability to recognize an environmental problem when they see one, and (2) *policy evaluation skills*, that is, the ability to evaluate proposed environmental policies. One would hope that students would also acquire certain rudimentary policy formulation skills, that is, the ability to formulate a policy solution to an identified problem; however, this does not strike me as a reasonable goal for an undergraduate curriculum.

These two analytical skills presume certain knowledge that students must acquire. These skills presume knowledge regarding environmental risk and risk management. Students need to be able to distinguish real problems from pseudo-problems and more serious from less serious problems; they need to be able to distinguish policy solutions that promise to solve the problem without creating even more intractable problems. These skills presume knowledge of the policy process within which solutions to environmental problems are crafted. Students need to understand, for example, how this process works and the roles that the key actors (e.g., interest groups, the press, legislators, and administrators) play in the process. These skills presume a knowledge of ethical theory, on the basis of which the equity, fairness, and ethical acceptability of proposed policies will be evaluated. The relevance and importance of the first two sorts of knowledge should be self-evident, so I will focus briefly on the third one.

Ethical theory bears on public policy, including environmental policy, not as a goal of such policy but rather as a constraint upon acceptable policy or as a criterion for being the right policy: Good public policy must, among other things, be an ethically acceptable, if not the ethically best, policy. Of course, enacted policy is not always good policy; in particular, enacted policy is not always ethically acceptable. Yet, even if ethical considerations are not decisive (and they often are not), they do play an important role in the policy process, especially in a political system like our own, where political power is widely dispersed among groups with competing interests and where accepted political ideology holds that exercises of political power should conform to ethical norms. Policies that are perceived by a significant portion of the public to be ethically unacceptable must survive in a hostile environment in which adoption, funding, and implementation are achieved only at great political cost to those who support the policy. Perceived ethical unacceptability often has a way of becoming political infeasibility.

Knowledge of ethical theory is crucial to an understanding of the policy, indeed the political, process in the United States, precisely because of the diversity of our ethical perspectives. Much, perhaps most, of the policy debate in the United States is ethical in focus, by which I mean that it concerns questions of rights, obligations, entitlement, just dessert, fairness, equity, and so on. If students are to understand and participate in this debate, they need an adequate grounding in ethical theory; they also need to know how and where ethical theory bears on the policy process. One may wonder how our forefathers were able to get by so well with so little formal education in ethical theory. The answer is that they lived in a much more homogeneous society, one in which political power was held by those who shared a single ethical perspective.

What students need to understand for the present purposes is the content and justification of the ethical theories that dominate U.S. political life. In particular, they understand the two dominant types of substantive ethical theory: rights-based and consequentialist ethics. They need to understand the two dominant strains or tendencies within rights-based ethics, libertarianism and egalitarianism, and the dominant type of consequentialist ethics, utilitarianism. They must also understand the crucial role that procedural ethics plays within our constitutional political system in resolving the deep ethical conflicts that pit consequentialists against libertarians and egalitarians. Students must also understand the contractarian justifications of these theories that proponents typically offer.

Once students understand the character of these ethical theories, they will be in a position to understand the ethical conflicts that often arise in the course of the policy process: Rights-based theorists focus on the question of the ethical acceptability of the structure of legal rights and obligations that the policy imposes on those affected by the policy, asking whether this structure is compatible with the dictates of their ethical theory; consequentialists, by contrast, focus on the question of the ethical acceptability of the policy's performance or consequences, not for each individual taken singly but for society as a whole. Given these different ethical perspectives, there will often be conflict and disagreement regarding the ethical acceptability of policies. These disagreements are not of a sort that scientific or technical information can resolve, because the disagreement has nothing to do with empirical fact; they can only be resolved procedurally or, failing that, by an exercise of raw power by the stronger of the parties.

Consider, for example, the increasingly frequent "water wars" in the Southwest that pit irrigation farmers against municipal and industrial users. Farmers argue that they should be permitted continued use of water at historical levels by virtue of their rights to the water; municipal and industrial users argue that they should be permitted use of this same water by virtue of the fact that their use of this water will maximize social welfare. There need be no factual dispute that scientific or technical expertise can adjudicate; the dispute here is normative in character: Should society allocate this resource to farmers, or should it allocate it to municipal and industrial users?

Means

The knowledge that the foregoing goals presumes (including knowledge of production agricultural practices and systems, farm economy, and farm structure; environmental risk and risk management; the policy process; and ethical theories) is generally available within colleges of agriculture or within the liberal arts colleges associated with colleges of agriculture. But this knowledge is rarely "packaged" in a way that focuses on environmental issues; more seriously, courses that provide one or another facet of this required knowledge rarely have as a curriculum goal the development of either the understanding or the analytical skills described above. If, by good fortune, a student happened to take courses that covered the relevant knowledge, it is unlikely that this student would succeed in integrating this knowledge. The interconnections are simply too complex and subtle. What is needed is a curriculum (1) that packages this knowledge in a way that focuses on, or at the very least calls attention to, its import for environmental issues, and (2) that provides some structure or mechanism for both integrating this knowledge and developing the requisite analytical skills. Changes of the first sort involve modifications, sometimes quite minor, to

179

existing courses; changes of the second sort, however, require the development of new courses whose goal is to provide this integration and to develop these analytical skills.

The particular way in which these courses are integrated within an existing curriculum will, of course, depend on local circumstances. Within Cook College, the land-grant college of Rutgers University, these courses constitute part of the college's general graduation requirement. Students must take at least 2 of the 21 courses offered (see box). I have included in the Appendix to this chapter the syllabus of one of these courses, Environmental Ethics, which I teach. That course is representative of many of the 21 courses. It has as its goals providing students with (1) an understanding of the ethical issues surrounding environmental policy and (2) the analytical skills to make evaluations of the ethical acceptability of proposed environmental policies. Sections A and B are intended to motivate and justify a course on ethics and environmental policy

---

Courses Constituting Part of Cook College's General Graduation Requirements

Perspectives on Agriculture and the Environment
Social and Ecological Aspects of Modern Agriculture
Human Responses to Chemicals in the Environment
A Systems Approach to Environmental and Agricultural Issues
Environmental Issues in the United States
Natural Hazards
Introduction to Systems Thinking and the Systems Approach
Economics of World Food Problems
Economic Growth, Man, and Environment
Elements of Environmental Pollution
Conservation of Natural Resources
Health and Environment in America
Population, Resources, and Environment
Energy and Society
Urban Society and Environment
Culture and the Environment
Politics of Environmental Issues
Social Responses to Environmental Problems
Social and Ecological Aspects of Health and Disease
Conservation Ecology
Environmental Ethics

(environmental ethics, for short). Section C, the theoretical heart of the course, examines the policy process, develops an analytical framework for policy analysis, introduces in some detail the major ethical theories that figure in public policy debate, and then discusses how ethics fits into the policy process, both as a determinant of policy and as an analytical tool. These first three sections occupy about half of the semester. Sections D through H are basically case study applications and illustrations of the theoretical framework introduced in section C. These latter sections of the course provide students an opportunity both to develop their analytical skills on real environmental problems and to learn something about these problems.

There is, of course, more to curriculum development than course development. Securing faculty agreement on proposed modifications to college graduation requirements is often difficult, especially within a college of agriculture, where the various disciplines are often clamoring for as much of a student's course time as they can get. There is also the difficulty of administering a curriculum of the sort envisioned. If the courses that constitute the curriculum remain under the administrative egis of the disciplines (or departments) that provide the faculty that staff them, it is difficult to ensure that the content of these courses actually furthers the curriculum goals described above. If, on the other hand, these courses are administered and controlled by the college, for example, by a standing subcommittee of the college's curriculum committee, one gains the leverage necessary to ensure that courses have the proper content, but only at the price of constant administrative oversight.

Support

Three sorts of support are crucial to the successful development of an environmentally sensitive curriculum. First, there must be adequate institutional support within the college. Institutional inertia and "turf management" problems make change and innovation difficult. More importantly, perhaps, faculty often are decidedly unenthusiastic about curriculum reform for the simple reason that they are rarely rewarded for the time they spend on curriculum development. These problems can be overcome to some extent by appropriate changes to the reward structures within the college. But the key to faculty involvement, especially in larger research-oriented universities, is research opportunity: The curriculum under development must open up opportunities for new research.

The requirement that proposed curricula generate research opportunities relates directly to a second sort of support that is crucial to the development of an environmentally responsible curriculum: There must be adequate research support to foster the development of needed curriculum materials. Courses that support the curriculum goals described above are difficult to develop and teach. Assembling and mastering the background materials necessary to design such a course are very time-consuming. Finding and assembling course materials suitable for student use are even more difficult and time-consuming. One invariably ends up having to assemble a photocopied collection of readings borrowed from a number of different sources. Many faculty, especially in less research-oriented institutions, simply do not have either the time or the access to information to undertake such a task. The desired curriculum changes will therefore come to those institutions only when the appropriate textbooks, anthologies, and research monographs are published. So, to the extent that one hopes to see curriculum changes take hold nationwide, there is going to have to be adequate research support.

Finally, the successful development of an environmentally responsible curriculum in colleges of agriculture requires what might be dubbed adequate extension support. The curriculum needs input from the field to ensure that it addresses real problems and equips students with the skills to handle these problems. The traditional land-grant concept of extension, research, and teaching is no less appropriate in environmental science than it is in agricultural science. Each of these three components is essential to the development of an environmentally responsible curriculum.

References

National Research Council. 1989. Alternative Agriculture. Washington, D.C.: National Academy Press.

Peters, M. D., ed. 1975. Letter to John Jay, August 23, 1785. P. 384 in The Portable Thomas Jefferson. New York: The Viking Press.

Appendix

Syllabus and Readings for the Course Environmental Ethics

There are any number of serious *environmental problems* that deserve public consideration and action, for example, acid rain, habitat destruction, species extinction, groundwater contamination, overpopulation, siltation of rivers, and deforestation. Environmental ethics is not concerned with these problems per se; rather, it is concerned with the *ethical issues* associated with these problems (e.g., our obligations, if any, to preserve endangered species and equitable distribution of natural resources). A single environmental problem may raise more than one ethical issue; the same ethical issue may be raised by more than one environmental problem.

The syllabus for the course is structured in terms of ethical issues; specific environmental problems and their attendant public policy initiatives are used to illustrate these issues.

A. What Are the Proper Goals of an Environmental Policy?

Environmentalists are sharply divided on the question of the proper goals of environmental policy: On the one side are the "preservationists," who argue that we have an obligation to preserve the natural environment in its "natural" state, while on the other side are the "conservationists," who argue that we have a right, perhaps even an obligation, to make optimal use of environmental/natural resources.

What one takes to be the proper goals of such policy influences both what one takes to be the salient ethical issues and what one takes to be the appropriate tools and techniques for dealing with issues that one recognizes. Preservationists, for example, have the task of explaining the source and scope of our supposed obligation to preserve the natural environment; conservationists, on the other hand, must answer such questions as for whose benefit should environmental resources be managed, and what counts as "optimal use"?

> *Readings:* Taylor, P. W. 1981. The ethics of respect for nature. Environmental Ethics 3:197–218.
> Baxter, W. 1986. People or penguins? Pp. 214–218 in People, Penguins, and Plastic Trees, D. VanDeVeer and C. Pierce, eds. Belmont, Calif.: Wadsworth.

B. How Ethical Issues Arise: The Coercive Nature of Public Policy

Many of the ethical issues raised by environmental problems stem from the simple fact that effective environmental policies are invariably coercive: environmental protection demands forms of social cooperation that can often be secured only by coercive means. One of the fundamental questions of environmental ethics asks under what circumstances is coercion justified.

> *Reading:* Schelling, T. 1978. On the ecology of micromotives. In Micromotives and Behavior. New York: W. W. Norton.

C. Ethics and Public Policy: An Analytical Framework

The complexity of the ethical issues raised by various environmental problems underscores the need for a coherent analytical framework from which we can assess and evaluate the ethical acceptability of public policies proposed to mitigate these problems.

In this section we begin with public policy (Ch. 2), asking what is public policy, defining the policy process, and developing a framework for analyzing public policy. We then turn to ethical theory (Ch. 3), first developing the notion of a "social contract," then considering rights-based ethical theories (libertarianism and egalitarianism), consequentialist ethical theories (utilitarianism), and proceduralist ethical theories. We next consider (Ch. 4) how ethics and policy fit together, discussing how ethical theory bears on public policy, how ethics figures in the policy process, rights-based (structure-focused) policy evaluation, consequentialist (performance-focused) evaluation (including cost-benefit analysis), and evaluative schemes that attempt to accommodate both rights and consequences.

> *Reading:* Thompson, P., R. J. Matthews, and E. O. van Ravenswaay.
> In press. Public Policy, Ethics, and Agriculture. New York: Macmillan
> (Chapters 2–4).

D. *Human Obligations to Wildlife*

Most environmental policy decisions have a significant impact on wildlife, typically through their impact on wildlife habitat. In this section we will examine the questions of what ethical obligations, if any, we have regarding wildlife (either as individuals or as species) and how such obligations as we do have are to be weighed against conflicting obligations we may have to other humans. We will consider critically the widely held assumption that the animal liberation movement is a close ally of the environmental movement.

> *Readings:* Sagoff, M. 1984. Animal liberation and environmental
> ethics: Bad marriage, quick divorce. Osgood Hall Law Journal
> 22:297–307.
> Gunn, A. S. 1984. Preserving rare species. Pp. 289–335 in Earthbound: New Essays in Environmental Ethics, T. Regan, ed. New
> York: Random House.

E. *Equity: What Counts as a Just Distribution of Environmental Resources?*

Many environmental problems have to do with the depletion of various natural resources. Policies that aim to regulate resource utilization face a basic ethical issue: How are competing claims on scarce environmental/natural resources to be adjudicated? What counts as a just distribution of these resources?

> *Readings:* Thompson, P., R. J. Matthews, and E. O. van Ravenswaay.
> In press. Public Policy, Ethics, and Agriculture. New York: Macmillan
> (Chapter 6).

Freeman, M. 1986. The ethical basis of the economic view of the environment. Pp. 218–227 in People, Penguins, and Plastic Trees, D. VanDeVeer and C. Pierce, eds. Belmont, Calif.: Wadsworth.

F. *Environmental Health and Safety: Acceptable Risk*

In matters of human health and safety, as elsewhere, there is no free lunch. Decreased risk comes only at a price. A fundamental ethical issue in environmental ethics asks about the levels of risk that we *should* find acceptable.

Readings: Portney, P., ed. 1978. Toxic substance policy. In U.S. Environmental Policy. Baltimore: Johns Hopkins University Press.
Johnson, D. 1986. The ethical dimensions of acceptable risk in food safety. Agriculture and Human Values III:171–179.
Shue, H. 1986. Food additives and 'minority rights': Carcinogens and children. Agriculture and Human Values III:191–200.

G. *International Problems*

Many environmental problems are global or at least international in character: Their solution requires international action. These problems raise issues regarding the source of any ethical obligations of nations to cooperate in these international actions as well as issues regarding the basis of any assessment of the equity of proposed solutions to these problems.

H. *Future Generations*

Environmental policy typically has consequences for future generations. In this section we examine the question of our obligations, if any, to future generations. Our examination considers such complicating factors as the contingency of future generations (i.e., it is to some extent up to us whether there will be future generations), uncertainty as to their precise needs, etc. We also examine the ethical import of the fact that certain necessities for human life are seemingly nonrenewable and irreplaceable.

RAPPORTEUR'S SUMMARY

Robert J. Matthews is a professor of philosophy and an environmental ethicist. He approached the topic of an environmentally sensitive curriculum from that perspective. He described an environmentally sensitive curriculum as one that will (1) develop in students the ability to recognize that a problem exists in agriculture as a result of environmental insults caused by several agricultural practices, (2) present to students knowledge of current problems and policies designed to mitigate these problems, and (3) develop in students analytical skills that will permit them not only to recognize problems but to evaluate them as well. He stated further that policy development activities should be included in ethics courses.

The discussion focused on how ethics subject matter should be incorporated into the curriculum. The consensus was that it should be accomplished through new integrating ethics courses and through portions (modules) of existing courses such as those offered in policy analysis, business and management, environmental education, and agricultural production, among others. It was stressed that an effective program requires not only coordination but faculty development workshops and other related activities as well.

Course content was discussed next. Matthews promoted the theme that students must be able to recognize ethical issues and how they fit into the policy process. In order to do this, they must understand policy processes and the two primary ethical theories used in the policy process: rights-based and consequentialist. Rights-based theories focus on individual rights, whereas consequentialist theories focus on maximizing social welfare. The knowledge base that these courses or modules require is usually available in all colleges of agriculture.

How to get faculty interested and involved in developing and participating in an environmentally sensitive curriculum received considerable discussion. Major themes included the following: (1) the need to have an effective reward system so that faculty are encouraged to participate rather than be penalized "because it took away from their research productivity," which itself is a good research area for some; (2) administrators must create a positive environment; (3) faculty development activities are essential; and (4) major collegewide curriculum renewal and revitalization efforts would foster many of these changes. One participant asked, "Why do faculty add units of many kinds to their courses and not others, such as environment and ethics?" The only answer suggested was that the units that are added are probably the result of one's own research activities.

The availability of course and faculty development materials was also discussed. A good source of case materials is the journal *Agriculture and Human Values* (Richard P. Haynes, Editor, University of Florida, Gainesville). Forthcoming by Macmillan is a book entitled *Public Policy, Ethics, and Agriculture* by P. Thompson, R. J. Matthews, and E. O. van Ravenswaay (Macmillan, New York). Two faculty development workshops on ethical aspects of food, agriculture, and natural resources were offered by the National Agriculture and Natural Resources Curriculum Project in June 1987 at the University of Kentucky. Several copies of the workshop materials are still available from Richard H. Merritt, the project director (send $25.00 to Department of Horticulture, Cook College, Rutgers University, P.O. Box 231, New Brunswick, NJ 08903). Another paper, "Integrating Agricultural and Environmental Studies in Colleges of Agriculture and Natural Resources," by Richard H. Merritt was commissioned by the congressional Office of Technology Assessment in 1989 and is available from the Superintendent of Documents, U.S. Government Printing Office, Washington, DC 20402. Other resources such as syllabi are available, as are results of courses concerned with agriculture, natural resource, and environmental studies now being offered in U.S. colleges and universities. Courses are offered at Texas A&M University; Cornell University; the University of Florida; California Polytechnic State University, San Luis Obispo; the University of Maryland; and Rutgers University, among others.

22

Breaking Traditions in Curriculum Design

C. Eugene Allen

Diana G. Oblinger, Rapporteur

In this discussion on breaking traditions in curriculum design, I plan to share some observations, raise some questions, and challenge us to think about some improved ways of designing and delivering a curriculum that is appropriate for the challenges of this decade. It is not the purpose of this chapter to provide a literature review on curriculum development.

This is a decade that has acquired many labels that are relevant to undergraduate education. Some of the following terms have been used to describe this era: the information age, the global and international era, the biological age, the decade of the undergraduate, and the environmental decade. Each of these labels implies that there are educational needs to be addressed. When agriculture passed from the labor to the mechanical era and then from the mechanical to the chemical era, it was necessary to make numerous adjustments to the agricultural curriculum. This decade, with its multiple advances and challenges, will tax our abilities to adequately prepare undergraduates for the complexities of the world in which they will live and work. The recognition of this fact and what we plan to do about it are the basic reasons for the national conference.

Bringing About Change

Observations at my own institution and others have led me to conclude that bringing about a major change in courses and curriculum at the undergraduate level is similar to confronting death, but unfortunately, in the curriculum area, even the seriously ill frequently hang on and achieve another birthday. If this is as commonplace as I suspect it is across many disciplines and colleges, then the problem of adequately addressing undergraduate education is made even more serious. When the world around us is rapidly evolving, it seems appropriate that our curricula must do the same.

Kenneth Christiansen of The Defiance College (Defiance, Ohio) has brought a perspective to curriculum change that seems helpful (Christiansen, 1988, 1989). In working with curriculum change and studying the feelings that were a part of it, he concluded that curriculum change is an "emotional as well as an intellectual process." As such, the emotional feelings are part of the normal reactions to grief over the loss of familiar ways. Christiansen (1988, 1989) advises that administrators working with the emotional part of curriculum change can assist the process by understanding the three stages of "grief work" described by Eric Lindemann (1944). These have been summarized (Christiansen, 1988, 1989) as follows:

1. Give up the past. Form a realistic view of what was lost by talking with another about the conflicts and pain of losing the past.

2. Build a realistic picture of the present. Understand that nothing is definite or enduring. Change is natural. Curriculum must change as times, knowledge, students, and needs change.

3. Form new relationships for the future. Focus ahead. Decide how personal, professional, and institutional needs will be met.

I have found this linkage to the grieving process to be useful in more fully understanding some of my own successes and failures in changing what I do or in working with course and curriculum changes either as a faculty member or a college dean. With this in mind, it should not be surprising that there are deep emotions tied to the curriculum because of ownership and what has become accepted as "the way it is done." In this sense, the changing of curriculum really is challenging a tradition with the idea that something different could be better. Such a challenge can easily put many faculty in a very defensive mode. In 1984, when I was interviewing for the position of dean of the College of Agriculture, I was asked what I would do if I had one wish as a new dean. My response was that I would have a button that could eliminate all courses and the college curriculum each decade. The reason for this response was to be able to do something meaningful in redesigning the curriculum and that this could best be done when faculty are in an offensive rather than a defensive mode. My assumption was that if there was no curriculum to defend, then everyone would be more likely to contribute to the vision for the new curriculum. After becoming dean, this concept became known as the

"Dean's Sunset Wish," and from this evolved Project Sunrise in the College of Agriculture at the University of Minnesota. The term *sunrise* was chosen to depict a new day or a new curriculum. Project Sunrise was funded by the Kellogg Foundation and resulted in some significant changes in curriculum, how faculty teach, and how the college's majors are structured. For example, there is much more emphasis on active learning, decision-case studies, interdisciplinary approaches, the need for breadth in undergraduate education, and the way that student advising is done. (Additional details about this project can be obtained by writing to the Dean, College of Agriculture, University of Minnesota, St. Paul, MN 55108.)

Curriculum Content Issues

The Project Sunrise experience and other experiences as a faculty member, administrator, and external program reviewer at other institutions have led me to the following observations and conclusions about undergraduate curricula, majors, faculty attitudes, and education needs in general. The first of these addresses issues relevant to the content of the curriculum.

• Too many faculty at research universities view the needs of the undergraduate primarily from a discipline point of view rather than recognizing the equally important aspects of interdisciplinary education. Undergraduates should be viewed as future alumni who will pursue multiple careers in their lifetimes, rather than as future professors who devote their entire careers to the narrow confines of a disciplinary area or subdiscipline. Only a small percentage of undergraduates will work in the narrow confines of a discipline for most or all of their careers.

• In too many departments and colleges, the primary changes that occur in courses and the curriculum are determined by who is retiring and who is hired. With the rapid expansion in new information, this is an increasingly unacceptable way to shape the curriculum. There is a need to more widely recognize that everything cannot be taught in an undergraduate program, but that certain concepts and principles must be taught as part of an appropriate foundation for present and future learning that is a part of how we define an "educated person." Among these "musts" I would include as examples the areas of communication skills, problem solving, cross-cultural understanding, important disciplinary concepts, and sufficient grounding in liberal arts and interdisciplinary courses to serve as a foundation for further personal or professional growth by the individual.

• The liberal arts and general education requirements of the curriculum should not be viewed just as requirements but as ways that truly enhance the intellectual growth of the individual. These courses should include not only the important aspects of the social, physical, and biological sciences, as well as the humanities, but also interdisciplinary courses that address major societal or world issues. In the latter regard, this is a natural way to improve relevancy and to involve faculty from professional and liberal arts colleges in jointly taught interdisciplinary classes. With some creative efforts by faculty, fewer students would view these courses as "just requirements toward a degree," and it would also serve to demonstrate the necessity of multiple considerations in addressing complex, real-world issues.

• The courses that may be most critical to setting the stage for an undergraduate education are not necessarily the introductory-level courses for a number of departments or disciplines. In other words, the sum of a series of disciplinary courses is not, in my view, equal, for general education purposes, to a few well-structured interdisciplinary courses that integrate material from a variety of disciplines. For example, an interdisciplinary course(s) on world food and hunger taught in an integrated way by faculty from different disciplines and taken by students from many majors should be a very different experience from the sum of disciplinary courses on introduction to food science, field crop production, and economic aspects of food distribution. These disciplinary courses may also be important to the curriculum, but in general, they would not be appropriate for the broader integrated understanding that many people believe is increasingly important.

• Improving the communications skills of undergraduates in all areas is a very high priority of many employers, and it is a responsibility that must be visible in many parts of the curriculum rather than just in a few courses offered by the English, speech, or rhetoric departments. Improving our communications skills comes from repetition that has the benefit of a helpful critique. When the curriculum does not emphasize communications and computer skills across the curriculum or in many courses, it seriously reduces the chances for improvement through practice. Likewise, when faculty require no essay test questions or give full credit when students use sloppy grammar, they are reinforcing poor communications habits. Research papers, oral presentations in class, student classroom discussions and debates, organizational activities, and laboratory or field trip reports are some of the ways that communications skills can be given some attention beyond the few credits that may be required in the curriculum. Finally, it goes without saying that any undergraduate today who is not at least semicomfortable with using a computer at the time of graduation will probably be at an immediate disadvantage.

• In general, our educational system from the elementary to the doctoral levels places a high premium on individual accomplishments. However, we increasingly hear that there is a need for employees to work as teams or in cooperative ventures that accomplish an organizational goal. This need should not be lost on how we teach or how we structure some classroom or laboratory activities. Some group projects can be built into the curriculum and can range from relatively simple to much more complex and time-consuming tasks or projects. There is a need for appropriate structuring and monitoring, but when they are done well, group projects can be a very positive addition and a stimulating and different kind of learning experience.

• There needs to be more integration of significant issues into the regular courses of the curriculum. Examples from issues like water quality, food safety, animal welfare, global markets, climatic effects, government policies, ethics, and the impacts of new technologies can frequently be integrated into regular courses as highly relevant considerations, even though these subjects may be entire courses by themselves. Such a practice helps to tie courses together and to demonstrate that most significant issues cannot be solved by simple answers.

From these examples, I have tried to address some curriculum issues that pertain to content and philosophy. In considering such issues, one should always be mindful of questions like the following:

• What are the highest priorities and goals for students in this curriculum?
• How can this curriculum be improved to meet the lifelong personal and professional learning needs of the students who are enrolled?
• What is the feedback from alumni and the employers of our graduates about this curriculum?

Process Items in Curriculum Revision

Next, I will address some issues that seem relevant to the process of bringing about curriculum change. We can have the best ideas around, but if we cannot get them incorporated or implemented, then they can only be discussed rather than tested. Too often, good ideas are never tested because the process for implementation and sustaining the change is inadequate for the challenges that must be addressed.

• Major curriculum change is sometimes approached in a revolutionary rather than an evolutionary way. My sense is that although revolutionary change may be needed, there is a significant chance of failure if a sufficient number of faculty have not bought into the changes in a sufficient way to sustain them. Two points can be made. First, getting people to buy in takes time, and saving time at the beginning may destine the curriculum change for failure either in the short or long term. Second, fewer revolutionary changes are necessary if a meaningful evolutionary process is in place and is operating each year. A revolutionary process is usually necessitated when meaningful evolution in curriculum is absent. In this situation, even small curriculum changes may be difficult. Furthermore, it is difficult to imagine an administrator or administrative team that can change the curriculum in the absence of significant faculty support. The key is how one gets the faculty to buy into the need for change, determine the process to be used, and develop or take ownership of the concepts that are to be incorporated into the curriculum.

• Creative group thinking and brainstorming about many issues, including undergraduate curriculum and education, are affected by the people, the surroundings, the timing in relation to other issues, and the time allowed for the discussion. It is for this reason that the department or college conference room seems to be such an ineffective place to achieve creative group thinking, at least in the beginning of the process. In this regard, I believe that a significant part of the success of Project Sunrise must be attributed to a few very productive joint faculty and administrative retreats that were held off campus with the specific purpose of discussing various dimensions of this project. These retreats were a critical part of bringing many people "on board" and developing a collective vision of where we were going.

• Picking the right people to lead the curriculum change effort is a critical first step. Even though there may be a great temptation for the leadership on such a project to come out of the dean's office, I believe that there is a more ideal model. There is always a significant risk that if the leadership is from the dean's office, it will become the "dean's project" rather than a more widely adopted "college curriculum project." It should not be difficult for the dean to pick key and respected faculty, a few interested students or recent alumni, and one or two key administrators to serve on the curriculum revision committee. The dean and the dean's office can then serve by providing a vision for revision, by inspiring and supporting the committee so that this vision expands throughout the college, and then by serving to challenge the committee and other faculty to achieve certain goals. The dean must also be a cheerleader for the change and set sufficiently high standards so that what is done makes a real improvement for the customer, which, of course, is the student. In bringing this about, the following thoughts may be useful. Remember that:

193

1. Most faculty are very proud and, as such, will not follow someone on a committee whom they do not respect. Therefore, this committee should consist of the most respected faculty who are involved in undergraduate teaching.

2. Although many faculty do not like to be told by their peers or administrators that it is time for some change, they are frequently more willing to listen to respected alumni and key employers of their students. The input and interest of such key groups should be sought early in the process.

3. The dean or whoever appoints the curriculum revision committee must be an avid supporter of the project through words and action. If the committee is appointed, given their charge, and basically told to report back when they are done, you can almost guarantee that nothing significant will happen except the probable wasting of much time in committee meetings.

• There is great need for faculty to feel that the time they spend working on curriculum change is not wasted. Remember that it would be unusual if many of them were not grieving and asking themselves, "What is in this for me?" A good administrator should be sensitive to these feelings and provide assistance and answers to these kinds of questions through actions, support, and enthusiasm for the project. When it is time for annual evaluations, the faculty who have made important contributions should receive appropriate praise and salary adjustments.

The collective vision for what is needed and what can realistically be achieved in curriculum revision for any unit is very dependent on the process that is used. Curriculum revision by its very nature requires that faculty give up some of the old curriculum, assist in creating a new curriculum, and then deliver the new curriculum to the students. Thus, whereas the administrators are theoretically in charge of distributing financial resources, for all practical purposes the faculty are in near complete control of the curriculum as givers, creators, and distributors. This means that the process for bringing about a change in the curriculum is more dependent upon leadership and less dependent upon the allocation of financial resources, as sometimes happens in bringing about a change in a research program. With this in mind, it should come as no surprise that it is sometimes very difficult to make significant changes in the curriculum. With regard to the process for curriculum change, the following questions may be helpful:

• Who are the key faculty who can provide the leadership for this change? What are their rewards for doing a good job?

• How can faculty come to be owners and believers in curriculum change?

• What is in this proposed change for students, faculty, and the college or unit?

• What are the challenges and disincentives to bringing about significant curriculum change?

Conclusions

I hope that this discussion has challenged the reader to think about some of the process and content considerations involved in curriculum revisions. I would like to conclude by sharing a few quotes from a sobering and thoughtful special report from The Carnegie Foundation for the Advancement of Teaching by Ernest L. Boyer (1990). I have chosen the following:

• The 1990s may well come to be remembered as the decade of the undergraduate in American higher education.

• What activities of the professoriate are most highly prized?

• At no time in our history has the need been greater for connecting the work of the academy to the social and environmental challenges beyond the campus.

• Disciplines have become increasingly divided.

• The educational experience of students frequently lacks coherence.

• Should some members of the professoriat be thought of primarily as researchers, and others as teachers?

• Designing new courses and participating in curricular innovations are examples of yet another type of professional work deserving recognition.

These quotations serve to emphasize some of the points that need to be considered in designing a curriculum and providing appropriate incentives for the faculty who create and deliver the curriculum. If this is to be the decade of the undergraduate, there is much that remains to be done with the ongoing process for curriculum evolution and the real and perceived values assigned to teaching compared with those assigned to research.

References

Boyer, E. L. 1990. Scholarship Reconsidered: Priorities of the Professoriate. Princeton, N.J.: The Carnegie Foundation for the Advancement of Teaching.

Christiansen, K. 1988. Core for anyone; on coping creatively with the needs of a faculty that is undergoing major curriculum changes. Paper presented at the Association of Integrated Studies Meeting, Arlington, Tex., October 1988. (Available from K. Christiansen, The Defiance College, Defiance, Ohio.)

Christiansen, K. 1989. Professional growth through curriculum development; coping with the pain of change. Paper presented at the Freshman Year Experience Conference, Cincinnati, Ohio, November 10, 1989. (Available from K. Christiansen, The Defiance College, Defiance, Ohio.)

Lindemann, E. 1944. Symptomatology and management of acute grief. American Journal of Psychiatry 101:141–145.

RAPPORTEUR'S SUMMARY

Context

As a basis for the discussions, Eugene Allen described Project Sunrise at the University of Minnesota. Their process of curriculum revision spanned 3 years, during which time 17 departmental majors were reduced to 11 interdepartmental ones. The process involved a coalition of faculty, the dean, and the vice president.

In developing a mindset toward curriculum change, the experiences at The Defiance College were reviewed. Curriculum change should be considered both an intellectual as well as an emotional process. To cope with the emotional side, we were reminded of the death and grieving process.

The central role of the faculty is a key element in curriculum change. Because the curriculum "belongs" to the faculty, it is their tradition(s) that is being challenged. They are the ones being asked to make changes. Curriculum tends to be a reinforcing process; that is, it is repeated year after year. Research, on the other hand, is constantly changing. This tends to make it more difficult to accept challenges to and changes in the curriculum.

Process

Faculty must take the lead in curriculum change. It is best to select respected faculty leaders. Consider only one representative from a department. This makes it easier for faculty to see the points of view of faculty in other departments. The administration should be involved as "cheerleaders." Recall that in the deliberations, the question of "What is in this for me?" must be answered either directly or indirectly.

The group process is extremely important in facilitating curriculum change. It is the group, not an individual, that develops the vision that will be articulated to the rest of the faculty. To encourage the development of effective groups, consider using retreats to begin the process. The following is an important reminder: developing group cohesiveness and formulating a plan will require significant time. Most colleges invest 2 to 3 years in curriculum change.

To foster the impetus for change, it may be useful to bring in outside groups or individuals. Employers and alumni can provide useful feedback on the preparedness of graduates and suggest programmatic changes. The techniques used include focus groups and "executives in residence."

In implementing curriculum change, provision should be made for faculty development. Most curriculum changes require the creation of a new awareness in faculty, the acquisition of new skills, and the need for alterations in the current modes of faculty-student interactions or classroom presentations.

Universities that are willing to discuss their processes with others include (but are not limited to) the University of Minnesota, the University of Hawaii, the University of Nebraska, and the University of Wisconsin.

Content

The amount of disciplinary content in revised curricula may appear to be reduced. However, recall the frequency with which professionals change careers. The focus of the undergraduate curriculum should be on developing well-rounded professionals rather than teaching future professors. The content of degree programs should be based more on the principles and skills that will serve graduates over a 40-year professional life rather than on faculty research topics or the current mix of faculty expertise. Historically, colleges of agriculture have tended to offer courses based on what faculty are interested in rather than what the students need.

While revising curricula, attention should be paid to the contribution that colleges of agriculture can make to nonmajors. The general educational value of a course on, for example, world food and hunger, taught in an interdisciplinary mode, can be substantial.

Core competencies can be woven into courses in colleges of agriculture. Components of writing, speaking, the development of teamwork, discussions of ethical issues, and the incorporation of real-world examples will enhance the quality of graduates from colleges of agriculture.

Criteria for Assessing Success

In assessing the success of curriculum changes, two measures are suggested: (1) the attitude toward teaching and (2) the change in graduates, as perceived by the students as well as by employers.

Key Questions

Throughout the curriculum change process, it is important to recall some key questions:

- Who are the key faculty?
- What are the rewards for doing a good job of curriculum change?
- How can faculty be owners of the process of change?
- What are the challenges and disincentives for change?
- What will the curriculum be?
- How will the curriculum be designed?
- Who will deliver the new curriculum?

23

Changing the Image of Agriculture Through Curriculum Innovation

Jo Handelsman

Jerry A. Cherry, Rapporteur

The last decade was an exciting time to be part of the agricultural community, which developed technological innovations of unparalleled drama. In the 1980s we witnessed events as unprecedented as the construction of the first transgenic animal, the use of remote-sensing systems for soil mapping, and the field testing of the first genetically engineered plant. This was also the decade of a new awareness and protectiveness toward our environment. From this awareness grew a new land ethic that began to align environmentalism with agricultural production. We replaced the vision of the soil as a growth medium for high-yielding corn with a regard for soil as a cherished resource and an integral part of our delicate landscape. It was a decade of renewal and discovery.

Despite these rich, attractive images and ideas that are the reality of agriculture in the late twentieth century, agriculture as a field of study is dogged by conservative, dusty, and dull images. It is regarded as a field of old-fashioned science and traditional technology practiced with wanton disregard for the environment. Sadly, students in our universities are more likely to associate agriculture with pictures of dark-suited, austere, nineteenth-century professors and one-horse plows than with casually dressed, twentieth-century molecular biologists and computer terminals. Even rarer in students' minds is an image of sophisticated teams of farmers, environmentalists, and agricultural scientists developing farming strategies that are friendly to the environment. The misconceptions about modern agricultural science must be due in part to the image that we in agriculture project. We are, and should be, proud of our rich history and our tremendous successes, but we must not promote our past to the exclusion of the fulfilling future that is unfolding.

I am often struck by the contrast between displays in the lobbies of buildings on the University of Wisconsin campus. The agricultural production departments often display rusty farm equipment and sepia-toned photographs of a bygone era. In contrast, the basic science departments dazzle visitors with shiny high-tech equipment or advertisements about study programs and recent research discoveries. Similarly, milestones in production departments are often marked with historical reviews paying homage to the past, while our neighbors in the basic sciences hold scientific symposia that honor their histories by teaching students how the past contributed to their current greatness. It is no wonder that our enrollments are dropping even as those in our sister departments in the life sciences are holding steady.

The content and pedagogy in many of our courses suggest that agriculture's projected image is indicative of a deeper problem. I believe that the need for change is surpassed only by the opportunity for change. We can, and must, treasure and teach about our past, but we must not stop there. We must show our students how our past has contributed to the present and the future, which is rich with potential for change and improvement. By so doing, we will project our strengths, improve our image, and attract students to our courses. Of greater importance, our courses will be current, substantive, and interesting. Students who find their way to these courses will be rewarded with a rich, intellectually fulfilling experience. I propose four challenges that, if they are met, will help to dispel the image of agricultural studies as insular and outdated. Meeting these challenges will potentially attract to courses in agriculture students from diverse social backgrounds and academic disciplines. We have the opportunity to attract students with the high-energy atmosphere that exists in agriculture today and to send those students to their chosen professions with a knowledge of agriculture's rich history and its challenging future. If we educate a larger audience in agricultural courses during the 4 years of university education, then we will generate a society that is more educated in the science and issues of agriculture.

I suggest that we challenge ourselves to find ways to (1) develop basic science courses that use examples from the agricultural sciences, (2) develop courses that explore the interface between society and agriculture, (3) develop nontraditional pedagogy that results in the active involvement of diverse perspectives, and (4) promote diversity among our teachers and students. Our challenges are as follows.

First, we need to find mechanisms to teach basic science courses that use examples from the agricultural sciences. Some of the most illustrative applications of basic science come from agriculture. Well-chosen examples will provide students with vivid, lasting images of how the basic science they have studied has been applied to problems of agricultural importance. The students will also appreciate the intimacy of the link between agriculture and mainstream basic science. If we place agricultural science in a broad scientific context, we are more likely to attract a wide spectrum of students.

This approach was tried in the Plant Pathology Department at the University of Wisconsin. We replaced a traditional course in bacterial pathogens of plants with a course that deals with the basic principles of host-parasite interactions and critical analysis of scientific papers by using examples from the plant pathology literature. After the approach changed, enrollment in the course tripled. Moreover, students were drawn from a greater range of departments, including those in basic biology, such as molecular biology and biochemistry.

Second, we must challenge ourselves to explore the interface between agriculture and other disciplines. By discovering the connections, we may find new insights into our own science and its societal context, and we may develop courses that attract students from a tremendous range of disciplines. Agriculture has always had a vast impact on economic development, societal structure, human relations, and demography. Today, society is faced with daunting choices about agriculture, and individuals are faced daily with personal choices that affect or are affected by agriculture. Courses that explore these impacts will contribute to developing an awareness of how agriculture shaped our history and how it affects our daily lives and decisions.

Numerous examples of courses that examine the interface of plant pathology and society are cropping up in universities across the country. These are popular with students majoring in history, environmental studies, journalism, economics, and education, because students are personally motivated by the relevance and timeliness of the topics to learn the science.

Third, I challenge us to be more creative in our choices of pedagogy. We must construct courses that actively involve students in the process of learning and teach them to synthesize information and solve problems. This is not only sound educationally but will also project the image of a field that is full of debate and deliberation and one that is open to new ideas.

I suggest three pedagogical approaches that stimulate student involvement. The first is the use of constructive conflict. This involves deliberately generating two sides to an issue and drawing the students into debate by requiring them to adopt a point of view and defend it. For this approach to be most successful, it is critical to demonstrate explicitly to the students the value of evaluating multiple points of view in learning to appreciate the complexity of a concept or issue. The second approach is the use of cooperative learning techniques, which require that the students work together noncompetitively toward a common goal. This approach is particularly useful for problem-solving exercises or for achieving consensus. Lastly, a pedagogical philosophy that is too broad to discuss fully in this context is feminist teaching. Briefly, the feminist approach is to develop a nonhierarchical classroom structure in which each member is an equal, respected contributor. In the feminist classroom, mutual respect is required and reinforced, providing an environment that is ideal for promoting creativity since it is conducive to taking intellectual risks. It can also be less threatening than the traditional classroom and therefore is an excellent environment for teaching analytical skills, since criticism is delivered and perceived as helpful and not personally threatening and disagreement is perceived as valuable and stimulating. Collectively, these and many other pedagogical tools can contribute to the construction of stimulating, dynamic learning atmospheres.

Fourth, we must attract a more diverse student body to dispel the image of insularity. I believe that pedagogy is the most important tool to do so, but there must be others. Agriculture's image will change if teachers of agriculture are perceived as soliciting, and even demanding, alternative viewpoints. Finally, classrooms that respect, value, and include the contributions of the students will be more likely to attract women and minorities, who often express a sense of alienation, exclusion, and disenfranchisement in the traditional classroom. I challenge us to find ways to interest women and minorities in agricultural science, to attract them to our courses, and to provide them with the positive environment, necessary stimulation, and sufficient feedback so that they feel that they are valued members of our educational community.

RAPPORTEUR'S SUMMARY

Jo Handelsman presented several challenges regarding curriculum innovation and the integration of agriculture-related courses into university curricula. The subsequent discussion emphasized the barriers and resistance to integrating agriscience into university curricula.

One participant expressed the belief that agriculture is best marketed as a basic science. Some participants felt that basic science fosters an exciting and challenging career image, while traditional discipline- and production-related courses project a comparatively mundane image. There was considerable discussion concerning the development of courses in colleges of agriculture that emphasize the basic sciences, with the objective of attracting students from outside colleges of agriculture and the cross-listing of courses in the agricultural sciences with the courses offered in basic science departments. Although basic science departments frequently object to and fail to accept such courses, several successful examples were provided. Somewhat surprisingly, there was a general consensus that faculty who teach the traditional agricultural sciences could offer greater resistance to such courses than faculty in basic science departments; several participants indicated that they had experienced greater resistance to innovation within departments and colleges than they had external resistance. There was also general agreement that colleges of agriculture lack aggressiveness in developing innovative courses and pursuing innovative teaching techniques.

Motivational factors were considered to be a barrier to curriculum revision. Effective revision can be handicapped by lack of incentives and rewards; certainly, the development of new courses is time-consuming and detracts from research efforts. Moreover, new courses sometimes fail to be implemented because of university bureaucracies. Cumbersome committee structures were criticized for being resistant to change. Problems with traditionalism and territorialism within departments and colleges were discussed. Some participants held the opinion that traditional departmental structures should be discontinued.

Internal and external perceptions of colleges of agriculture were considered an impediment to effective curriculum revision. It was emphasized that the role of colleges of agriculture differs at different institutions, but colleges of agriculture tend to project a poor image in comparison with the images of some other colleges within a university system. Both faculty and students were criticized for contributing to this problem. It was recognized that faculty and students within most colleges of agriculture tend to have comparatively traditional and conservative philosophies. There was general agreement that colleges of agriculture would benefit immeasurably from increased diversity.

Colleges of agriculture tend to suffer from insufficient representation of women, minorities, and others with diverse social and cultural backgrounds. Every effort to promote diversity within our system was encouraged.

In summary, the discussion group believed that it is feasible and proper to integrate agriscience and business into university curricula. Although external impediments to such innovation will frequently be encountered, internal resistance can be the greatest impediment.

Teaching Science as Inquiry

―――――――――――――

Paul H. Williams

Alvin L. Young, Rapporteur

Science and I

I sense,
I wonder;
What if I . . . ?
How can I . . . ?
I'm doing it!
I did it!
Wow!
How do I know . . . ?
How can I tell you I know?

Since the turn of the century, science and agriscience education across the United States has been reinforcing a dilemma of increasing magnitude and of its own making. Put simply, scientists and educators have preempted science from the public domain by substituting the knowledge created by scientific inquiry for the process of creative inquiry. Today, the domain of scientific inquiry largely resides within the increasingly remote arenas of the specialized disciplines, which are protected by academic rights of passage and secret languages.

In late-twentieth-century schools and colleges, science has become a "spectator activity" fueled by more sophisticated communications technology, in which knowledge has become a collectible and a commodity, where textbooks are like baseball cards, where somehow if you "bone up" on all the stats and facts you will pass through the portals of academe and find yourself on the playing field in the big league of science. *Nova* and National Geographic specials are equivalent to the most enticing hamburger and whole-grain cereal ads: Somehow they are supposed to be good for you, yet all they provide is a momentary sense of satisfaction, awe, or indignation. With the very latest innovations in interactive video communication technology, information vendors are poised to "Nintendo" our children and our teachers into even deeper and more subtle diversions from what the real game of science is all about.

Science begins when any human of any age who is curious about some phenomenon in the natural world around them begins to question and explore the relationships of the phenomenon to their understanding of their world. Science begins with an observation and a question and proceeds through a process of inquiry involving exploration and investigation, experimentation and analysis, and exposition and persuasion. That process engages the creative energy of the individual and leads to a deeper understanding, a sense of pleasure, and increased self-worth. Young children do science quite naturally: "Look what I found, Mom!"

Science, as it is taught in most schools across the United States, preempts teachers and students of the pleasures of creative inquiry by substituting the disciplinary content of knowledge for pleasureful and sometimes painful opportunities to discover for themselves why their world is the way it is.

To provide a basis for discussion on teaching science as inquiry, I prepared a few brief assertions and questions.

1. The process of inquiry is truncated when a correct answer is taught, given, or required.

2. Is there a general paradigm that characterizes the process of scientific inquiry as opposed to other forms of human inquiry?

3. Technological innovation can be a useful partner in the process of scientific inquiry.

4. Science should be a participatory activity, with both the teacher and the student engaging in the game.

5. How can scientific inquiry best be taught?

6. What is the appropriate balance between knowledge content and the inquiry process?

7. What are the constraints to participating in inquiry-based teaching?

RAPPORTEUR'S SUMMARY

Most of us who participated in the conference had someone, likely a teacher, who gave us a sense of excitement about the field of science. From that initial spark of excitement we developed careers in science, and for most of us, many areas of science are even more exciting today, as we have seen the advancements of an incredible array of new technological tools for inquiring into science. Yet, my 15-year-old son, a freshman in high school, reports to me that the textbook and science teacher are "boring" and that he frequently goes through a great deal of agony to learn something simple. The question posed for this discussion group was most appropriate; namely, "How do we teach science in a way that captures and returns the students' interest?" Simply put, "How do we teach science as inquiry?"

Science is the process of discovery. But the conditions conducive for discovery are not always present. The following four impediments were identified.

1. *Relevancy.* In the daily life of the student, the process of discovery seems to have nothing to do with the process of science. Too frequently, to the student and general public, there is no relationship between the food in our supermarkets and the research conducted by our land-grant universities.

2. *Tunnel vision.* Today, we are struggling with students who see the end of the tunnel (college) as a route to getting jobs. Science courses are seen only as requirements in traversing that tunnel, that is, blocks to be checked on a scorecard.

3. *Socialization.* Science courses in our institutions frequently become socialized. They are not designed with a creativity factor in mind. Rather, they are structured to meet the student's needs, that is, helping them to check off the blocks on their scorecards. Hence, the result is aimless lectures and multiple-choice tests.

4. *Documentation.* The way in which we report science "turns off" students. We prepare articles that must meet rigid publication requirements—a very conservative process that leaves our journal articles dull. We publish hundreds of journals that are read by a small number of people and certainly few, if any, students. Moreover, there are few science writers who can make science exciting to the reading public.

In view of these impediments, how do we teach science in a manner that is pleasurable, exciting, and educational? Participants made the following suggestions.

1. *Balance content and process.* We must find a balance between the knowledge content of science and the inquiry process. Science courses must be more than just a collection of knowledge. In teaching genetics, I can remember the excitement of the students when we combined the knowledge of meiosis with mathematics, enabling us to predict and understand the ratios in the segregation of genetic traits in our laboratory studies of fruit flies.

2. *A participatory process.* Teaching science should be an interactive situation between the student and the teacher, both partici-

pating together in conducting an experiment. The close relationship between a graduate student and the major professor exemplifies a situation in which science is best taught.

3. *The use of technology.* Technological innovation can be a very useful partner in the process of scientific inquiry. For example, the simplistic setup required for tissue culturing allows this powerful technology to be readily used in the laboratory as a technique for problem solving and scientific inquiry.

4. *Independent study.* At the undergraduate level, independent study provides an excellent way to teach science. The selection of the project and the followthrough of the subsequent steps to its successful completion can instill in students a sense of excitement about science.

5. *Communication and persuasion.* Teaching students in a peer relationship situation provides a cooperative learning mode, because the student must communicate accurately (stating the hypothesis), demonstrate results (testing the hypothesis), and persuade peers that the question was answered (proving or disproving the hypothesis).

6. *Alternatives to "correct" answers.* Young children do science naturally. They formulate ideas. They make observations, and they make conclusions. By the time a child reaches high school, most of the creative nature is gone. The process of inquiry is truncated when we demand a correct answer. We do not encourage deviations from our correct answer, nor do we encourage other interpretations to fit the hypothesis. As teachers, we must be willing to say that we do not know whether this is the correct and only answer. We must encourage students to evaluate many solutions to a problem.

7. *Reporting science.* The most difficult tasks in the process of inquiry are interpreting and reporting the observations. How we communicate among peers through our journals frequently limits or restricts how we communicate with the public. We should, perhaps as a minimum, require our students and ourselves to write summaries of our reports in a manner that an outsider could read and understand.

In summary, to teachers of science, the lessons from this discussion are that the science we teach should be relevant to the life of the student, that it contains a balance between the knowledge content and the inquiry process, that it should be a participatory activity, with both the teacher and the student engaging in the game, and that the tools of technology can make it an innovative process. As we develop our science curriculum, we must remember that our goals are the growth and maturation of our students so that they are able to understand and apply the scientific process.

207

Emphasizing the Social Sciences and Humanities

Paul B. Thompson

William P. Browne, Rapporteur

Many authors writing on the future of undergraduate education in colleges that have historically had their roots in agriculture have stressed the need for a broad view of society and culture. Virtually no one has called for more specialized training in the applied sciences. Given this background, much of the work that might have been expected in a chapter emphasizing the social sciences and humanities has already been done. What is more, the completion of the Social Science Agricultural Agenda Project (Johnson et al., 1991) has produced a wealth of material for those who wish to find a more detailed discussion of the topic. Three key points need to be made with respect to the role of the social sciences and humanities in the education of agricultural and natural resources professionals.

1. Social science and humanities courses play a dual role in undergraduate education. They are a part of core undergraduate education, but certain social science and humanities topics have special relevance for the careers that agricultural and natural resources professionals will pursue in the twenty-first century.

2. The social sciences and humanities are the only disciplines within the university that are equipped to help students understand the way that an increasingly urban population will perceive food and fiber as well as environmental issues. If agricultural and natural resources professionals are to be effective in their careers, they must be prepared to listen and reply to the concerns and desires voiced by people with little life experience or formal education in the production of food and fiber or in the management of natural resources.

3. The capacity for targeted social science and humanities education on topics and skills crucial to future professionals in agriculture and natural resources is both low and poorly organized. This area has been among the most neglected by faculty administrators in both the agricultural and the environmental sciences and in the liberal arts.

The balance of this chapter takes up each of these three points in turn. The importance of a broad education is taken as a given; there is no further discussion of it here.

The Difference Between Core and Targeted Education in the Social Sciences and Humanities

Many of the contributors to this volume on professional education for undergraduates have stressed the need for broadening the curriculum of agricultural and natural resources professionals. They have cited the need for courses in communications and foreign languages and a core area of knowledge about culture and society. There is nothing special here that relates to agriculture or natural resources. Business, engineering, premedicine, and prelaw students have this need. It is a real need that must be acknowledged, but recognition of this need should not influence the curriculum reform effort in agriculture and natural resources to focus on developing a core curriculum.

Agricultural and natural resources professionals face special problems of communication, ethical decision making, and interpreting and managing human activities, problems that are unique to the types of careers they will follow and to the kind of science they will apply. The social sciences and humanities can and must be incorporated into their training in such a way as to target educational efforts on the acquisition of knowledge and skill that is specifically relevant to these problems. Some of the specific topics that should be targeted are discussed below.

The single most important point that must be recognized, however, is that emphasis upon core humanities and social science courses does not substitute for the targeted education of special social science and humanities topics of particular importance to agriculture and natural resources. Programs of core education that stress "great books" or a unified approach to understanding society and civilization through the study of art, literature, or history are quite likely to exclude these special topics in a systematic and deliberate way. If these special topics are not introduced into the education of agricultural and natural resources students at the upper division or graduate level, the core education movement to emphasize the social sciences and humanities will, in fact, deprive future professionals of the social science and humanities knowledge that is most crucial to their effectiveness. Note that this is not meant to oppose the core movement, for all students need it. It is crucial to see that there are two tracks for discussing the role of social sciences and humanities in the education of agricultural and natural resources professionals. One is the core needed by everyone; the other is the targeted areas needed by agricultural and natural resources professionals. Everyone needs the core. Agricultural and natural resources students need something else, preferably in addition to the core. What they need is the topic that follows.

The Content of Targeted Areas in Social Science and the Humanities

U.S. agriculture enjoyed a reputation for success during much of its tenure, but the most recent decade has been one of criticism and rethinking of agricultural priorities (Danbom, 1986; Johnson, 1984; Kirkendall, 1987). One important group of critics, associated with the 1972 "Pound" committee of the National Research Council (1972, 1975), stressed the scientific quality and efficiency of agricultural research, but the more noted critics have focused on the social goals that contemporary agricultural production techniques (whether implicitly or intentionally) have tended to serve (Berry, 1977; Doyle, 1985; Fox, 1986; Hightower, 1975; Jackson, 1980; Schell, 1984). There is an extraordinary range of concerns and complaints expressed in the writings of this latter group of critics, and many different client groups are alleged to have been ill-served. A theme common to most criticisms, however, is that agricultural leaders have, de facto or by design, pursued a goal of maximizing the productive efficiency of the U.S. farm. Critics allege that it is the persistent search for greater yields by the U.S. Department of Agriculture (USDA) land-grant system that is the wellspring of problems for U.S. agriculture.

The views of the critics were reinforced by a legal finding in November 1987, when California Rural Legal Assistance (CRLA) won a judgment against the University of California (UC). CRLA claimed that producers who have aggressively sought a competitive edge have benefited disproportionately from publicly funded agricultural science, at the expense of small farms and the farm labor that had been displaced by the resulting technological changes. The court found that UC had negligently failed to assess whether research to develop a mechanical tomato harvester would have an adverse impact upon UC's legislatively mandated small-farm clients (Bishop, 1987). Although the judgment against UC was later reversed, the court action and the press coverage it engendered indicate the seriousness of criticisms raised against agriculture.

Other critics have cited what they perceive to be negative impacts on environmental quality, the poor or oppressed peoples of developing nations, consumer health, and even the welfare of farm animals. The common theme is that the USDA land-grant system's service to the (increasingly) large farm has lowered the price and improved the availability of food and fiber at the expense of other social goals. The ongoing public controversy over recombinant bovine somatotropin (rBST) brings many of these elements together in a single case. Although the scientific evidence indicates that rBST, which is produced through a genetic engineering process, is a safe and efficient technology for dairy production, use of the technology has been stifled by a coalition of small dairy producers and animal welfare and consumer advocates. This coalition has, to date, succeeded in keeping milk produced by cows treated with rBST from reaching consumers, largely by raising concerns about the safety and quality of the product (Burkhardt, 1991).

The tomato harvester and rBST cases are but two of many that complicate the current context of agricultural production, research, distribution, and consumption. Others include the questions of field testing engineered organisms, determining the acceptable risks associated with agricultural chemical residues in food, examining our commitment to the development of agriculture in less developed countries in light of domestic farm interests, preserving genetic diversity, and environmental and public health regulations as barriers to trade in agricultural products. These are among the most difficult of a long list of topics that, more broadly, include world hunger, environmental quality, animal welfare, and the traditional agrarian philosophy of farming as a way of life.

Today's agricultural leaders, not to mention today's citizens, need a more sophisticated understanding of the food and fiber system. They need to appreciate the social, ethical, and cultural values that seem to surface only in a crisis situation. Although it is not clear that crises such as the banning of Alar or California's "Big Green" referendum (which would have banned a large number of agricultural chemicals and addressed other environmental concerns) can be anticipated with any degree of confidence, a deeper understanding of the social and cultural forces that are operative during crisis situations will help agricultural leaders make more effective and responsible responses to public concerns, when they arise.

Traditional agricultural education in the humanities and social sciences has stressed economic management of farms and agribusinesses as well as rural community development. Although there will be a continuing need for this education for a percentage of students being educated in traditional agricultural programs, there is an even greater need for education on how society beyond the farm sector relates to and perceives agriculture. Future professionals will need to know how to do a better job of producing the products that urban consumers want. They will need to know how to manage their professional responsibilities in a manner consistent with the public interest. They will need to know why people who do not have a farm or rural background might find certain practices or ways of speaking arrogant, insensitive, or otherwise objectionable. They will need to know how to listen to a new constituency.

Present-day farm, food industry, and environmental leaders are keenly aware of the fact that the consumer-oriented and political decisions that define the framework for agricultural and natural resources management are made by people who have no life experience or formal education in agriculture or the environmental sciences. The general public did not grow up on farms and has little or no experience with the productive use of natural resources. Many lack life experiences even in the recreational use of nature. The dietary choices and opinions on regulatory issues of most Americans are not informed by knowledge of principles for evaluating and comparing risks, nor are they informed by information on the contribution of existing farming and management practices to food availability, economic growth, and the provision of other human needs. Although we should work to promote better public understanding of agricultural and natural resources management, we must plan the education of the next generation of leaders on the assumption that this situation is not likely to improve. It would be tragically foolish to think that the public at large will assume the personal costs needed to understand the scientific and production-based opinions of those who make their life in agriculture, the food industry, and resource management. It is agricultural and natural resources professionals who must bear the responsibility to communicate with the public. This means that leaders must understand public opinion on its own terms. Put simply, the mountain will not come to us; we must go to the mountain.

Specifically, this means that undergraduates need to study how nature and natural resources are perceived in U.S. society. They need to learn alternative philosophical approaches to the measurement and acceptability of risk. They need to be taught how scientific advances such as biotechnology are received by different sectors of the public and whether public reactions are based on political and financial interests or on moral and religious concerns. There is a need for graduate and undergraduate training in communications strategies that do not alienate nonscientific, nonfarm audiences. There should be undergraduate courses that take up the politics of the policy process and the histories and organizational structures of groups (commodity organizations, environmental or animal welfare activist organizations, consumer groups, etc.) that influence agricultural and natural resources practices. Journalism departments should offer courses on how the news media decides which stories to cover and of the norms and institutions that structure the coverage

212

of science and technology. Future professionals who possess knowledge and skills in each of these areas will be far more effective in conducting and promoting research, product development, marketing, management, and policy change than those who do not. The volumes of the Social Science Agricultural Agenda Project (Johnson et al., 1991) document the subject matter for such targeted course work in an exhaustive manner.

Yet, it is clear that the current social science capacities within colleges that have their roots in agriculture have little capacity to educate undergraduates (much less to do research and extension) on these issues. This lack of capacity is partly organizational. Marketing and resource economics courses in agricultural economics are targeted to advanced majors. A similar situation holds for course work on social psychology, development theory, and cultural analysis that might be offered by sociologists and anthropologists with appointments in colleges of agriculture and natural resources. Such organization does little to serve the broad educational needs of undergraduates. A more serious problem exists with respect to the educational needs that derive from political science, communications, journalism, literature, and philosophy. With the exception of agricultural communications or journalism programs aimed primarily at training tomorrow's agricultural press, capacities for educating undergraduates on these topics within colleges of agriculture and natural resources are practically nonexistent. What, then, are the barriers to reform?

Barriers to Emphasizing the Social Sciences and Humanities

During the 1980s, a number of journals and professional societies emerged to support research on the broad social and ethical issues that have spawned conflict, controversy, and the need for better communications between agricultural and natural resources professionals and the public at large. *Agriculture and Human Values, Issues in Science and Technology,* and *The Journal of Agricultural Ethics* have joined traditional outlets such as *Science, Environment,* plus other environmental journals and many monographs and anthologies devoted to agricultural and natural resources issues. The Agriculture, Food and Human Values Society (University of Florida, Gainesville) has, at the time of this writing, over 700 members. The basic knowledge for meeting teaching needs exists to a far greater extent than it did in 1980. Although there will be a continuing need for research and publication in the areas of agriculture, environment, and societal values, the lack of models and materials can no longer be accepted as an excuse for not offering educational opportunities to undergraduates.

Curriculum reform is a faculty-based process: decisions about what is to be taught are ultimately made by faculty. Since faculty are not likely to place themselves in a position in which they are expected to teach subjects and methods in which they perceive themselves to have little expertise, the existing capacity of agricultural faculty places a severe constraint upon the direction and extent of change in the agricultural curriculum. Curriculum reform has largely meant that standard courses in the plant and animal sciences substitute the study of gene transfer and computer technology for the study of mechanical and chemical technology (and it is a partial substitution at that). In some instances, courses that approach production problems in terms of cropping systems or farm management are being replaced by course work that takes an even narrower approach, generally assuming that an ability to identify the genetic basis of economically valuable traits need be the only item in agricultural scientists' tool kit for the coming generation. Even in the social sciences, the response has often favored replicating the management and computer systems curricula currently offered in colleges of business.

To the extent that agricultural curriculum efforts have tended to increase the capacity for exploiting discoveries in molecular biology and computer technology, one can argue that they have entirely failed to respond to the needs outlined in this chapter and may, in fact, constitute an abandonment of the special historical mission of agricultural education. The emphasis on technology responds to declining enrollments by introducing a curriculum that undergraduates perceive to offer training in marketable skills. Biotechnology and computers are not peculiarly suited to agricultural and natural resources management, however, and the new corps of undergraduates correctly perceive that their ability to exploit their technical training in no way depends upon a sophisticated or reflective understanding of the food and fiber and the natural resources system. The state of agricultural education's ability to investigate and disseminate a comprehensive and unified vision of food systems in modern society has, if anything, been damaged rather than improved by curriculum reform efforts that stress the hot new technologies.

Faculty may also have thought that such a stress would be consistent with the existing research and educational capacities of agricultural and natural resources faculty. In fact, however, the move has been accomplished by importing new faculty whose training and experience gave them no particular basis for loyalty to agricultural or natural resources management. In effect, the move has allowed the administrative structure and faculty lines of former colleges of agriculture to be captured by both students and young faculty who have no particular interest in or understanding of agriculture, natural resources, or the social systems that support them. This capture is, of course, partial; a majority of faculty and administrators still have traditional roots in agriculture and natural resources. However, the fact of this capture points us toward both problems and opportunities for colleges of agriculture to respond to the need for change in their curricula. New faculty (and students) will not only lack the capacity to deal with a broader notion of agriculture, they will also lack any reason to regard their educational mission as encompassing such broader concepts. It may soon be clear that enhancements in the direction of biotechnology and computers respond to declining undergraduate enrollments at the expense of the traditional land-grant mission, and that they leave food producers, natural resources managers, and rural communities without any educational organizations that are committed to the creation and dissemination of knowledge in support of their interests and ways of life.

On the other side of the equation, there is reason to doubt that the liberal arts disciplines can supply teaching expertise in the required areas on many land-grant university campuses. Although there are many individuals who have such expertise, they are not equally distributed throughout all universities. Liberal arts departments that have concentrated on achieving disciplinary expertise or quantitative skills in areas such as sociology, political science, history, philosophy, and literature are quite unlikely to have hired and promoted faculty members who specialize in science policy, environmental studies, risk issues, or science communication during the past decade. Simply inviting these departments to offer course work for agricultural and natural resources students would produce a disaster at some universities.

There are, of course, more familiar and mundane barriers to the advancement of the agricultural education system with a broad and sophisticated vision of the social, cultural, and ethical dimensions of the production and distribution processes, consumption patterns, and management possibilities for renewable resources. Tenure and promotion, opportunities for publication, and funding sources all readily come to mind. To a large extent, however, these more commonly cited barriers are all functions of the existing research and educational capacity within agricultural universities, since each is the result of expectations and values that are held by individuals who currently occupy faculty and administrative posts. Although it would be naive to ignore such barriers in promoting curriculum change, it would be equally naive to think that an effective effort to enhance an agricultural faculty's ability to integrate values issues into more technical subjects will not simultaneously improve the prospects for overcoming institutional barriers of this sort.

Opportunities for Change

There are also a number of institutions where innovation in the social sciences and humanities has been successfully targeted to the subject matter areas needed for agricultural and natural resources professionals. In each of the efforts where enhancement has succeeded, a four-stage process has been followed. First, there has been a careful effort of planning and coordinating a "core group" with members drawn from a variety of disciplines. The members of this group must agree to work together over a long period of time and must be willing to respect the integrity of other members. This is especially the case when the group must agree to disagree on some point, but must get on with the business of planning and coordinating larger activities. Second, there has been explicit attention to what might be called "market development" for the activity that is to be carried out. In the case of a group that produced a book on research policy at Texas A&M University (Thompson and Stout, 1991), this consisted of identifying faculty and administrators who would take time to participate in some of the workshops offered by the group. The third stage is the dissemination of information through workshops or conferences or workshops that, by outward appearance, resemble conventional academic activity. This is the most efficient way to disseminate ideas among a faculty that is used to the idea of attending conferences. Finally, there has been a follow-up phase in which faculty from the various disciplines maintain contact, sometimes in a systematic way by initiating more structured collaborative projects, but often by way of informal networking. These four stages are not necessarily a temporal succession; they represent levels of activity that can be pursued simultaneously. The possibility of enhancing capacity on ethics, social values, and food and fiber systems depends upon understanding how each stage presents tasks that must be accomplished if the goal is to be achieved.

Planning

All activities require planning, of course. What is special here is that the planning group includes people who must talk to one another across disciplines and who will be sensitive to the barriers that are imposed by jargon and the reigning values of people in different academic departments. This group must develop a rapport and must be willing to make a multiyear commitment, although the total number of hours required from each participant may be quite small.

Marketing

The effort required to find and prepare an audience for the primary product will depend upon many decisions that are made in the planning process. At a minimum, the success of the project requires preliminary networking and identification of agricultural and natural resources faculty members who will be supportive. The process also requires gestures from the administration that demonstrate its seriousness. The creation of positions and the expenditure of money may be painful, but there is no better way to communicate seriousness.

Transfer

Transfer is something that academic professionals know how to do; but organizing, advertising, administering, and presenting workshops consume both time and money. In this area, workshops should stress experiential learning techniques, case studies, and role-playing simulations for students. These approaches are essential if faculty who do not have disciplinary expertise in values studies are to teach values issues in the classroom. They are also the most effective educational techniques for students who are pointed toward careers in agriculture and business. Although learning modules that use these techniques must be constantly developed, updated, and refined, this is one area where the work of the past decade has put us in good stead to accomplish some dissemination of modules in the 1990s.

Follow-Up

Follow-up includes more networking to ensure that those who participate in workshops continue to receive information and support. There should be nationally coordinated efforts, so that faculty who are not trained in social science and humanities disciplines have continuing access to those who are. Like marketing, follow-up activities depend a great deal on the circumstances of particular individuals; so much of what must be done cannot be described in advance. There has been too little organized follow-up from previous agricultural and liberal arts projects or from curriculum development activities sponsored by the office of Higher Education Programs of USDA.

A commitment by the dozen best agricultural and natural resources universities to demonstrate progress in emphasizing the social sciences and humanities over the next decade would produce the change. If each institution committed two full-time-equivalent faculty members who were split primarily among faculty members with disciplinary training in political science, anthropology, history, philosophy, and communications, the cadre of professionals that would be created would be the spark for change. If funds were available to ensure that these professionals would have an opportunity to network with one another and with agricultural and natural resources professionals, change would be assured at those dozen institutions. A coordination of this effort through USDA's office of Higher Education Programs would ensure that other institutions could follow along at considerably lower cost. Although financial resources will be scarce during the coming decade, the commitment that is needed to ensure change is but a tiny fraction of the investment that has recently been made in moving toward biotechnology and computers. Emphasis on the broader social context of agriculture and natural resources should be regarded as an insurance premium paid to protect that investment from the kind of public reaction that overtook the nuclear power industry in the 1970s. From that perspective, a nationwide group of 24 full-time-equivalent faculty members with some supporting funds for networking and research seems like a small price to pay.

References

Berry, W. 1977. The Unsettling of America. San Francisco: Sierra Club Books.

Bishop, K. 1987. California U. told to change research to aid small farms. The New York Times, November 19, 1987, p. 13A.

Burkhardt, J. 1991. Ethics and Technical Change: The Case of BST. Discussion Paper 91.02. College Station, Tex.: Center for Biotechnology Policy and Ethics.

Danbom, D. B. 1986. Publicly sponsored agricultural research in the United States from an historical perspective. Pp. 142–162 in New Directions for Agriculture and Agricultural Research: Neglected Dimensions and Emerging Alternatives, K. A. Dahlberg, ed. Totowa, N.J.: Rowman and Allanheld.

Doyle, J. 1985. Altered Harvest. New York: Viking Penguin.

Fox, M. W. 1986. Agricide: The Hidden Crisis That Affects Us All. New York: Schocken.

Hightower, J. 1975. The case for the family farm. Food for People, Not for Profit, C. Lerza, and M. Jacobson, eds. New York: Ballantine.

Jackson, W. 1980. New Roots for Agriculture. San Francisco: Friends of the Earth.

Johnson, G. L. 1984. Academia Needs a New Covenant for Serving Agriculture. Mississippi State Agricultural and Forestry Experiment Station Special Publication. Mississippi State: Mississippi State Agricultural and Forestry Experiment Station.

Johnson, G. L., J. T. Bonen, and D. L. Fienup. 1991. Social Science Agricultural Agenda and Strategies. East Lansing: Michigan State University Press.

Kirkendall, R. S. 1987. Up to now: A history of American agriculture from Jefferson to revolution to crisis. Agriculture and Human Values 4(1):4–26.

National Research Council. 1972. Report of the Committee on Research Advisory to the U.S. Department of Agriculture. Washington, D.C.: National Academy of Sciences.

National Research Council. 1975. Agricultural Production Research Efficiency. Washington, D.C.: National Academy of Sciences.

Schell, O. 1984. Modern Meat. New York: Random House.

Thompson, P. B., and B. A. Stout, eds. 1991. Beyond the Large Farm: Ethics and Research Goals for Agriculture. Boulder, Colo.: Westview Press.

RAPPORTEUR'S SUMMARY

U.S. and world agriculture are beset by contemporary issues of environmental protection, energy applications, and nutritional standards. There is, no doubt, an element of the new in the way that these issues are addressed: New social values, new public policy claimants, and new solutions abound. Of course, little is ever truly new in the world of food and fiber production. Production agriculture, especially as embodied in the traditions of the land-grant system, has long debated stewardship practices, energy use as a production cost, and the quality of what is produced. So there exists a richness in what colleges of agriculture can offer to what many mistakenly see as new debates.

The varied remarks of the nearly 40 participants in this discussion group emphasized two points about this interface of old and new perspectives as these center on university curricula. First, colleges of agriculture, which are already financially strapped and which occupy a minority status in education, must charge their faculty and staffs with the task of being socially relevant to new land-grant constituents while not losing touch with the old ones. That is, production agriculture and food and fiber industries must be served while the contributions of other social forces are brought to bear on teaching, research, and extension. In essence, traditional colleges of agriculture need a regenerating boost or they will lose even more stature.

Second, and as a response to the first point, this boost may well come most effectively from work undertaken jointly by those in agriculture in partnership with those educated in the humanities and social sciences. Common agreement existed in the group on the need to incorporate philosophy, political science, history, and sociology into the core of some agricultural studies. In addition, students need to be able to write more persuasively, reason more soundly, appreciate more varying views, and see their work in a broader social and ethical context when they leave traditional agri-

219

cultural programs. The technical skills of science and methodology are by themselves insufficient.

The emphasis on these two points in the discussion brought to the group a clear understanding of the primary mission facing those who do curriculum revision. Before moving on to courses and content, they must decide what the undergraduate should be able to do and comprehend at the completion of a program of study. Only after deciding this can other disciplines be integrated and the level and mix of specially needed skills be determined. Moreover, the administrative and instructional arrangements for faculty and staff cooperation, which are to be used most efficiently in difficult financial times, await decisions on what we want students to be after they have moved beyond the undergraduate experiences we provide them.

Determining what we want students to be need not be done in the absence of cooperative ventures between agriculturalists and those in the humanities and social sciences, however. Joint planning is necessary. In addition, there are already a number of specialized appointments, multidisciplinary courses, and integrating experiences throughout the land-grant system. The results of these experiences should be evaluated and then reviewed carefully to determine whether students who shared them benefited appropriately. Did they better comprehend the position of agriculture in a changing world? Did they develop skills that they thought were useful in grappling with the increased expectations faced by those previously turned out into that world? Did they get jobs? Did they keep them?

As guidance to this evaluative review of past cooperative practices, four areas of questioning were suggested as being essential to curriculum redesign in each institution.

1. What degree of substantive literacy in agriculture must be linked to each undergraduate degree program or major? Must substance be understood, in a developmental sense, as this component of agricultural knowledge has developed over time?

2. To what extent are the technical and methodological skills of the instructional discipline that monitors each program or major necessary to the student's course of study? Is too much being attempted and is the student product too narrow?

3. Which broader skills, such as writing, are necessary to employ effectively the student's knowledge of agriculture and techniques of the monitoring discipline? Where in the university, in an instructional sense, are these skills housed? Can they be transferred from there?

4. Which types of broadening experiences, such as awareness of the public policy process, can contribute to the student's ability to apply successfully the things learned about agriculture from the

monitoring discipline? Who in the university, in an instructional sense, knows enough about the content of agriculture to bring these additionally useful experiences to students? Can cooperation across department or college lines be ensured?

When these four sets of questions about student needs and the university's ability to provide for them are answered, curriculum designers can move on to compare and contrast the benefits of numerous administrative approaches to organizing instruction: multidisciplinary programs or majors, interdisciplinary courses, joint faculty appointments, or even the creation of departments that have no disciplinary center.

Because resource allocations vary, as do the unique histories of each university, various land-grant colleges will probably select different administrative alternatives in producing their students of choice and, as a result, the instructional strategies needed to mold them. Given the variety of options and conditions, what is right for one state and its university could seldom be expected to be right for another. Thus, with an institutional review focused on the student product, the land-grant system should be expected to continue its diversification.

Teaching Agricultural Science as a System

Donald M. Vietor and Laurence D. Moore
C. Jerry Nelson, Rapporteur

Ten years ago, respondents to a survey by the National Higher Education Committee ranked "food and agricultural systems analysis" and "problem solving" high among course areas that were not adequately represented in agricultural curricula. The National Agriculture and Natural Resources Curriculum Project was organized under the direction of Richard H. Merritt of Cook College (Rutgers University) to respond to the National Higher Education Committee's assessment of curriculum needs. A task force of university faculty, the Systems Task Force, was organized in 1982 to develop curriculum materials and conduct workshops that would contribute to the teaching of systems analysis in colleges of agriculture.

The ideas about systems techniques and methodologies presented here reflect Donald Vietor's learning in the context of the Systems Task Force and associated workshops and Laurence Moore's experiences during his successful promotion of systems approaches to agricultural production problems in Virginia. Our objective is to present ideas and approaches to systems from our knowledge and experience that will stimulate interaction among those who would teach agricultural science as a system. In addition, the world view of systems approaches presented here, particularly the "soft systems," provided concepts, techniques, and models of inquiry that shaped the design of activities for involving the participants in this session.

Definitions

In order to teach agricultural science as a system, it is necessary to define *system*. The term *system* can be used to describe a set of elements or components that are connected together to form a whole (Checkland, 1981). These components function together in support of the objectives of the whole. This definition of system further stipulates that the properties of the whole emerge as a function of the connections and interactions among components. The emergent properties of the whole cannot be understood or explained by studying the components in isolation or apart from interactions with other components and the environment. An "agricultural system" can be perceived as comprising interacting biological and physical components that form a whole with emergent properties (Lowrance et al., 1984). Connecting any group of components together does not necessarily result in emergent properties for the whole; that is, it does not constitute a system (Rykiel, 1984). A description of operational units of agriculture as systems (Spedding, 1979), without consideration of emergent properties, is inconsistent with the definition of system submitted here.

A "systems approach" takes a broad view that concentrates on interactions among parts and on emergent properties of systems that are relevant to problematic situations (Checkland, 1981). The term *approach* describes a way of doing. Here, doing focuses on problems relevant to agriculture.

Models: Means or Ends?

The attention given to the development and evaluation of quantitative models within agricultural disciplines and journals can contribute to perceptions that techniques of simulation modeling and linear programming equate with systems analysis and systems approaches. Rapid progress has been made in the modeling and simulation of agricultural processes during the past 20 years. Models are available to simulate processes such as weather, hydrology, nutrient cycling and movement, tillage, soil erosion, soil temperature, and crop growth and development (Jones and Kiniry, 1986). Models can indicate where deficiencies in current scientific knowledge exist (Bawden et al., 1984). They can serve purposes of exploration, explanation, projection, and prediction (Rykiel, 1984).

Conceptual and quantitative modeling can be useful in the practice of reductionist science and technology development in agriculture. Mechanisms or technologies can be modeled apart from and in the context of higher levels of organization in support of hypotheses and experimental designs.

Since the age of Newton, reductionist science has contributed to verification of mechanisms and models through focused inquiry and experiments on selected parts of complex phenomena (Checkland, 1981). The integration of mechanisms into biophysical models, using the language of mathematics, can accomplish the purposes described by Rykiel (1984). These models can represent and convey the knowledge of those who built them. Yet, biophysical models may not meet the criteria set forth in the definition of system. Emergent properties may be absent. In addition, the model may be irrelevant to the current problems facing agriculture.

For example, models received attention during early stages of the learning and curriculum development activities of the Systems Task Force. Conceptual and quantitative models of different world views of the "agricultural system" that was developed by individuals in the group were considered among the potential materials for teaching systems approaches. Task force discussions revealed that each model of an agricultural system represented a simplified view of reality that was unique to its author and to the reality it represented. Moreover, most quantitative models in the published literature have been ignored by all except the model builder and have had relatively short lives (Rykiel, 1984). Using the definition of system presented earlier and the experience of the Systems Task Force, the notion equating systems approaches with modeling is inappropriate. It may be unrealistic to expect that agricultural science can be taught as a system through presentation and manipulation of published versions of biophysical models. Whose model, that is, whose system, will be used?

Applied Systems Analysis

The value of conceptual and quantitative models is best realized in the context of methodologies or processes for tackling problems and researching systems ideas. In general, agriculturalists are more concerned with real-world applications of systems ideas to solve problems in contexts ranging from farm to government policy levels than they are with studies of systems ideas for their own sake. Modeling is just one stage of systems approaches to problem solving in agriculture (Clayden et al., 1984). Applied systems analysis and the associated use of computer-processed models are most useful in settings in which goals can be specified, performance can be monitored, and implementation can be achieved. This quantitative approach evolved in the context of machine-based or hardware-dominated systems. The phrase *hard systems analysis* has been used to describe the approach which presupposes that a defined need exists, in the form of a perceived difference between a current and desired state, and that optimal solutions are both feasible and realistic goals for an analyst working to achieve the desired state (Checkland, 1981; Naughton, 1984).

Applications of hard systems include systems engineering and aids to decision making. Systems engineering is concerned with conceiving, designing, evaluating, and implementing a system of interacting components that meets a specified need (Naughton, 1984). For example, using quantitative models and simulation, a whole system of interacting components can be designed for optimizing the efficiency of alternative fuel production from crop biomass. Systems analysis can aid decision making through quantitative appraisals of the costs and consequences of alternative means of achieving the desired state or defined objective.

Systems ideas have been applied to aid producer decisions about stocking their pastures. Computer-aided decision making can help managers accomplish the objective of maximizing profit in an environment of changing costs for livestock and pasture production. A mathematical model can be constructed to describe the interdependence of stocking rate, animal- and pasture-related costs, and animal performance (Vietor et al., 1982). The model quantifies the trade-off among goals for maximizing performance per animal and per hectare and maximizing profit per hectare. What-if experiments that use the model can assist management of the stocking rate in support of the goal of maximizing the amount of profit per hectare as costs change.

Clayden and colleagues (1984) have described eight stages of a hard systems approach for achieving a desired goal. The first step is identifying and describing the problem to be solved and the existing system and environment. Second, the objectives of decisionmakers and the constraints are identified in relation to the problem or opportunity. Third, alternative routes for achieving objectives are generated and narrowed down to a set of the most feasible options. Next, measures of performance are established for optimization before the fifth stage of model construction. The models serve to predict outcomes when comparing options for achieving objectives. In the sixth stage, measures of performance are used to evaluate various routes to the objectives. In addition, the model itself is evaluated to determine whether it is representative of the real world. Unquantifiable objectives and constraints come into the picture at stages seven and eight, when the best options are chosen and implemented, respectively.

Experiential Learning

Initially, the members of the Systems Task Force lacked a common language and paradigm for learning and for thinking about and applying systems ideas. The diverse disciplines (human ecology, agricultural engineering, agronomy, agricultural economics, and social ecology) represented among the members were confounded with variations in individual approaches to problems that ranged from reductionism to holism. The fundamental epistemological and methodological differences among disciplines and individual scientists have made it very difficult to communicate and to reach agreement about the ways in which problematic situations should be approached and students should be taught (Buttel, 1985).

A model of learning (Kolb et al., 1979), not teaching, was the foundation of a common language and of paradigms for tackling problematic situations that were learned and shared among members of the Systems Task Force. This model of experiential learning illustrated the interplay between human experience and abstract thinking and the roles for both reflection and action. Human activities represented in models of reductionist approaches to science and technology development and of the steps of hard systems analysis (Bawden et al., 1984; Wilson and Morren, 1990) are analogous to those of experiential learning. Similarities notwithstanding, the activities or stages of reductionist approaches to science and technology and of applied systems analysis are practiced at different levels within a hierarchy of inquiry (Figure 26-1) (Bawden et al., 1984). Differences among scientists and among students with respect to the level of inquiry that each prefers within this hierarchy are potential sources of disagreement.

Reductionist scientists may argue that knowledge and methods for achieving the goals of agriculture will be advanced more through studies of mechanisms that function at the cellular or biochemical level than those that function at a systems level. Conversely, applied scientists and technologists who serve producers may view reductionist science as too narrowly focused and discipline oriented, emphasizing science without contributing to the knowledge base of modern farming (Bradshaw and Marquart, 1990). Exploration, practice, and discussion of this hierarchy of inquiry enables learners and problem solvers to assess and compare approaches to learning and improvement of problematic situations. The role of biophysical and systems models at each level of inquiry is illustrated above. A focus on a hierarchy of inquiry and learning activities may be more relevant to teaching agricultural science as a system than is a focus on a body of subject matter or on modeling techniques and models of agricultural systems.

Agricultural Science Under Fire

Once heralded as an example of human conquest over nature, technologies resulting from reductionist approaches to agricultural science are now under fire from critics. The persistent search for greater crop and livestock productivity supported by these technologies has been perceived by critics as a major source of problems facing American agriculture (Thompson, 1988). Both private and public agents of technology transfer have been influenced by the shift in emphasis from one of maximizing crop yields and pest

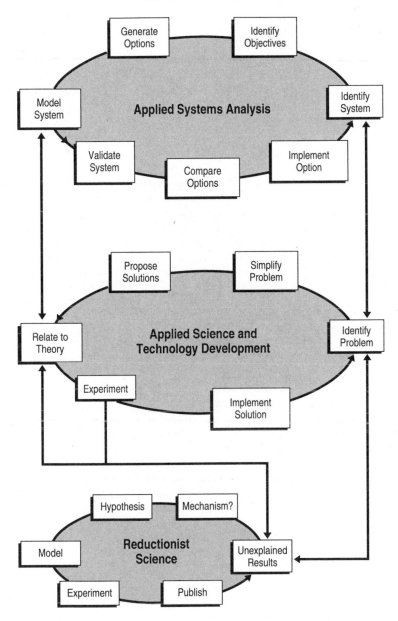

FIGURE 26-1 Conceptual model illustrating the relationship among applied systems analysis and reductionist approaches to science and technology development. Source: Bawden, R. J., R. D. Macadam, R. J. Packham, and I. Valentine. 1984. Systems thinking and practice in the education of agriculturalists. Agricultural Systems 13:205–225.

227

control to one of maximizing food safety and environmental protection (Bradshaw and Marquart, 1990). The time frame of the effort of the Systems Task Force coincided with this shift in public concern.

As this task force considered alternative approaches to inquiry and problem solving, the utility of reductionist approaches to science and technology development and of hard systems analysis were questioned, much as were the traditional goals of agriculture (maximizing productivity and profitability, optimizing production efficiency). Members of the task force, like others in agriculture, were forced to examine traditional ways of teaching, learning, and problem solving. What did agriculturalists need to do differently in practice and in the education of students to cope with the external forces confronting agriculture? Implicit in the title of this chapter, "Teaching Agricultural Science as a System," is the same question.

It was the perception of the Systems Task Force that information and technologies from levels of inquiry represented by applied systems analysis, applied science and technology development, and reductionist science (Figure 26-1) would not satisfy critics of agriculture as long as the goals and objectives of inquiry originated largely from agricultural scientists and their clientele. To date, information produced from these levels of inquiry has not answered accusations that the agricultural research system has failed to admit responsibility for problems arising from agricultural technologies and practices (Heichel, 1990).

The Systems Task Force was challenged to identify an approach to inquiry and problems that would prepare agricultural graduates to function in an environment of conflict over goals and objectives for agriculture. The approach would need to be useful when change was indicated, but the direction and means for change were problematic (Checkland, 1981; Holt and Schoorl, 1989). The goal-seeking nature of reductionist approaches and of applied systems analysis (Figure 26-1) appeared to be an incomplete representation of the range of human endeavor needed in agriculture. Then and now, a systems approach that goes beyond quantifying relationships among soil, plants, and animals is needed. Agricultural development includes relationships among people (producers, processors, and consumers), in addition to their natural and physical environment (Bawden, 1989).

Soft Systems Methodology

The ideas and methodology of soft systems offer an alternative to goal-seeking paradigms of applied systems analysis and reductionist science and technology. Previous applications of this methodology indicated that issues and concerns of participants in prob-

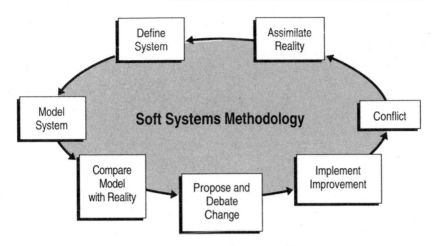

FIGURE 26-2 Conceptual model of stages of soft systems methodology. Source: Bawden, R. J., R. D. Macadam, R. J. Packham, and I. Valentine. 1984. Systems thinking and practice in the education of agriculturalists. Agricultural Systems 13:205–225.

lematic situations in agriculture, including critics as well as clientele, could be considered in determinations of what was problematic (Bawden et al., 1984; Macadam et al., 1990). During stages of "finding out" and of "debating proposals for improvement," the soft systems methodology facilitates self-conscious choices by participants. Those choices determine the purposes of learning and systems thinking. In contrast, an expert's preconceived notion of the agricultural system often determines what questions are asked and what is problematic in the paradigms for applied or hard systems analysis and reductionist approaches.

Using soft systems, the researcher and the researched, the consultant and the client, and the proponent and the critic work together in a dynamic relationship to identify goals or purposes while they collaborate to learn about the situation that they share (Bawden, 1989). The researcher or analyst serves as a facilitator without the pretenses of being completely objective, an expert, and detached from the problematic situation or opportunity. The techniques and methodology of soft systems facilitate consensus amidst the uncertainty present in complex situations (Checkland, 1981).

Soft systems can be modeled as a holistic approach to experiential learning (Figure 26-2) (Bawden et al., 1984). This methodology is organized into discernible stages and uses techniques that have evolved from both practice and theory (Checkland, 1981). The methodology can facilitate improvements in situations where there is conflict over what is problematic, including situations concerned

229

with the teaching of agricultural science. In addition, soft systems provide a conceptual framework for researching soft and applied systems methodologies themselves.

Mutually related judgments of reality and value can be part of the process (Vickers, 1968). While assimilating reality, a mutual appreciation of values among participants from within and from the environment of an agricultural situation can replace argument or conflict with dialogue. Systems of human activities that "could be" are defined and modeled to be relevant to the collective concerns of participants in a problematic situation (Wilson and Morren, 1990). A practical wisdom can arise from the collective concerns unique to each problematic situation as participants debate proposals for improvement (Vietor and Cralle, 1990).

Soft systems provide a more holistic or higher level in the hierarchy of inquiry (Figures 26-1 and 26-2). This methodology provides perspective and a clearer focus of inquiry for subtending, goal-oriented learning at levels of applied systems analysis, applied science and technology, and reductionist science (Bawden et al., 1984). Subtending levels provide insights for upper levels. The learner moves from level to level as each learning situation (i.e., problem or opportunity) requires.

Beyond Lectures and Expert Advice

If each of us reflects on the way that we were taught during our undergraduate years, we may recognize that the teacher was the principal learner in the classroom. Teachers, like scientists, determined the focus of inquiry through their choice of subject matter and related problems. The cognitive abilities of recall and comprehension were required of students to a much greater degree than were application, analysis, synthesis, and judgment. Teachers were similarly responsible for choosing and demonstrating those skills for manipulating plants, animals, soil, and environment that students should learn. Teachers were in control. Students were expected to integrate the knowledge and skills they gained from courses in science, rhetoric, mathematics, humanities, social sciences, and their own disciplines as they emerged into the professional environment after graduation. The relationship of student to teacher was not unlike that of clientele to the agricultural scientist.

Recently, agricultural consultants have expressed concern that agricultural research, the fruit of agricultural scientists, is too narrowly focused and discipline oriented, often emphasizing science and ignoring practice. Is an analogous criticism applicable to the teaching of agricultural science? Is the subject matter of agricultural science relevant to public concerns for human and environmental health and agricultural sustainability? Are students prepared to work in an environment marked by conflict among world views represented within and outside the farm gate?

The ideas and methodologies of experiential learning and soft systems can complement the propositional (learning for knowing) and practical (learning for doing) learning that have been emphasized in traditional approaches to teaching agricultural science (Bawden, 1989). Unlike propositional and practical learning, experiential learning depends on a dynamic interplay between sensory experiences of the world and mental abstractions (Bawden, 1989). The uniqueness of experiences, perceptions, and conceptual thinking for each learner suggests an approach that is learner rather than teacher centered. Learning of agricultural science, both cognitive and conative, will be motivated by the experiences of learners. What experiences are currently available to undergraduates in colleges of agriculture that motivate students to learn agricultural science?

Learner activities, both explicit and implicit, in the soft systems methodology (Figure 26-2) illustrate what can be done to cope with the complex issues facing agricultural science today. This holistic approach presents a role for the learner that differs from the purportedly detached and objective role of the scientist. The roles of the values and perceptions of the learner are acknowledged. Students conceptualize and learn with other players in agriculture in response to problems and opportunities unique to each new situation. Practice of the soft systems methodology in today's agriculture will stimulate students to seek and learn agricultural science that is relevant. Should one goal of curriculum reform be to teach agricultural science as a system? Or, should it be to encourage a systemic approach to inquiry that facilitates learning of agricultural science?

References

Bawden, R. J. 1989. Towards action researching systems. First International Action Research Symposium, March 20–23, 1989, Bardon, Queensland, Australia.

Bawden, R. J., R. D. Macadam, R. J. Packham, and I. Valentine. 1984. Systems thinking and practice in the education of agriculturalists. Agricultural Systems 13:205–225.

Bradshaw, D. E., and D. J. Marquart. 1990. New age professionals for a new agricultural age. Agrichemical Age (May):24–25.

Buttel, F. H. 1985. The land-grant system: A sociological perspective on value conflicts and ethical issues. Agriculture and Human Values 11:78–95.

Checkland, P. B. 1981. Systems Thinking, Systems Practice. Chichester, United Kingdom: Wiley.

Clayden, D., J. Hughes, L. Jones, and J. Tait. 1984. The Hard Systems Approach: Systems Models. Milton Keynes, United Kingdom: The Open University Press.

Heichel, G. H. 1990. Communicating the agricultural research agenda: Implications for policy. Journal of Production Agriculture 3:20–24.

Holt, J. E., and D. Schoorl. 1989. Putting ideas into practice. Agricultural Systems 130:155–171.

Jones, C. A., and J. R. Kiniry. 1986. CERES-Maize: A Simulation Model of Maize Growth and Development. College Station: Texas A&M University Press.

Kolb, D. A., L. M. Rubin, and J. M. McIntyre. 1979. Organizational Psychology. An Experiential Approach. Englewood Cliffs, N.J.: Prentice-Hall.

Lowrance, R., B. R. Stinner, and G. J. House. 1984. Agricultural Ecosystems: Uniting Concepts. New York: Wiley.

Macadam, R., I. Britton, D. Russell, and W. Potts. 1990. The use of soft systems methodology to improve the adoption of Australian cotton growers of the Siratac Computer-Based Crop Management System. Agricultural Systems 34:1–14.

Naughton, J. 1984. Soft Systems Analysis: An Introductory Guide. Milton Keynes, United Kingdom: The Open University Press.

Rykiel, E. J., Jr. 1984. Modeling agroecosystems: Lessons from ecology. Pp. 157–178 in Agricultural Ecosystems: Unifying Concepts, R. Lowrance, B. Stinner, and G. House, eds. New York: Wiley.

Spedding, C. R. W. 1979. An Introduction to Agricultural Systems. London: Applied Science Publishers.

Thompson, P. B. 1988. Ethical dilemmas in agriculture: the need for recognition and resolution. Agriculture and Human Values V:4–15.

Vickers, G. V. 1968. Value Systems and Social Process. New York: Basic Books.

Vietor, D. M., and H. T. Cralle. 1990. Comparison: Stage 5 of the soft systems approach. Systems Approaches for Improvement in Agriculture and Resource Management, K. Wilson and G. E. B. Morren, Jr., eds. New York: Macmillan.

Vietor, D. M., R. M. Rouquette, Jr., B. E. Conrad, and M. E. Riewe. 1982. Computer aided instruction: An economic analysis of pasture management. Journal of Agronomic Education 11:17–21.

Wilson, K., and G. E. B. Morren, Jr. 1990. Systems Approaches for Improvement in Agriculture and Resource Management. New York: Macmillan.

RAPPORTEUR'S SUMMARY

The use of a systems approach for teaching agricultural science was appropriately introduced by Laurence Moore and Donald Vietor. Moore reminded the participants of the discussion group that teaching covers a spectrum from disciplinary approaches, which are usually unilateral in terms of input, to holistic approaches, which depend on group interactions and problem solving. He effectively challenged the participants to think broadly in terms of the problem-solving method, and especially the use of a soft systems approach to education.

Many members of the discussion group were not fully acquainted with the hierarchy of systems approaches, and thus, it was defined in terms ranging from the reductionist approaches typical of those used by the practicing researcher to the open, participatory approaches involving students from a variety of disciplines. The four tiers of teaching or learning technology were described as follows:

1. The scientific method, which is a strongly reductionist approach. It is a well-accepted method for generating technology.

2. The application of technology, which expands the use of systems approaches. This requires an understanding of the technology and creative insight to visualize situations for its application.

3. The hard systems approach, which is frequently associated with mathematical models and model building. In this case, a desired change can be described and inputs or outputs can be calculated or understood. This approach has a great deal of quantifiable input and output, but the inputs and decisions are generally from one individual.

4. The soft systems approach, which is more conceptual and does not depend on mathematical models. It involves input from several individuals, often in a group setting, to achieve a desired outcome. The science base along with humanistic implications and social values are expressed and integrated into the outcome during the decision-making process.

One can visualize the hierarchy as (step 1) a scientist who determines that the yield of corn responds to nitrogen application because enzymes convert nitrate from the soil into ammonia and the amino acids that are assembled into the proteins needed for metabolism and growth. Others use that information (step 2) to learn that the efficiency of the response of corn is altered, depending on when and in what form the nitrogen fertilizer is applied, or that wheat yield is also increased by nitrogen application. The hard systems approach (step 3) would evaluate quantitatively the fertilization practice in terms of the nitrogen cycle, plant uptake, and soil losses as affected by crop rotation, dates of planting, and other agronomic practices. Specific goals from nitrogen application such as maximizing economic return, minimizing nitrate in groundwater, or decreasing weed infestation in subsequent crops can be evaluated mathematically by using the model. The soft systems approach (step 4) would add other dimensions; for example, how would the alteration of fertilization practices influence the local economy, the effectiveness of the school system, the quality of the public water supply, the abundance of wildlife, or the visual appearance of the landscape? Many of these latter outcomes are not quantifiably determined; rather, they are value judgments often made by individuals who are not directly involved with the nitrogen decision-making process. Thus, a soft systems analysis would be based, at least partially, on a broad range of inputs, albeit with variable strengths or impacts, regarding the correct nitrogen decision for the total system.

To introduce systems concepts to my undergraduate students I use the analogy of a partly cloudy day, that is, scattered cumulus clouds floating gently overhead while being surrounded by blue sky. Each cloud represents a cluster of knowledge or the technology of a discipline or subdiscipline. One of the objectives of a learner in problem solving is to read the clues or technical inputs contained in each of the several clouds of knowledge and then to integrate them, in effect to coalesce the clouds into a more dynamic set of interacting technologies. A major effort in learning is to be able to anticipate, determine, and evaluate the linkages, that is, the relationships between steps 1 and 2 above. The application of mathematical formulas to quantify the relationships moves us to step 3.

The other example I use is the spider web, with its intricate interwoven network of slender threads (or a set of elements that are connected together) that forms a whole. Pressing on one intersection of the web causes it to move, but every other intersection also moves, with the actual movement (impact) being dependent on the distance from the pressure point. The challenge for the students is to define the factors that are affected by a given decision (nitrogen fertilizer rate) and assigning each factor to a location relative to the pressure point. In a limited soft systems manner, the backgrounds and perspective of the individuals in the class are reflected through the selected (defined) input and outputs for the "decision" and, especially, the distance (relative strength) that each one is placed from the origin or pressure point within the web.

The steps in systems analysis are to analyze, synthesize, judge, and apply. On the basis of this sequence, the participants in the discussion session were divided into six groups for a discussion of the issues raised by Donald Vietor and Laurence Moore and their concerns about the systems approach, that is, analysis, or step 1. The factors reported back by one or more groups included the following:

1. The systems approach adds relevance to the reasons for why things are learned and, in fact, helps to define science in a broader perspective.

2. The systems approach provides a sense of input and leadership for students, and students develop confidence in problem solving, with more of a focus on group rather than individual decision making.

3. The systems approach helps to bridge the training and education relationship, especially the balance between gaining knowledge and understanding the approach to knowledge.

4. The systems approach may cause students to act more as generalists in their approach rather than as specialists, which would occur when they are focused through a specific discipline.

5. Less emphasis on a discipline may reduce the amount of cognitive material that can be covered in the curriculum, but not all courses would need to use systems approaches.

6. Students may be more prepared for the systems approach than are the faculty. More faculty time would be required to develop objectives and a format for teaching the systems approach than for traditional lectures.

7. Many classes are large, which may lead to compromises in teaching approaches, but most groups acknowledged that there are probably some innovative ways around this problem.

8. Faculty, in general, are reductionist. A challenge will be to find receptive faculty who can be motivated and rewarded for refocusing on the systems approach.

The groups then discussed what could be done to accomplish more systems approaches in student learning, that is, synthesis and application, or step 2. The factors reported back included the following:

1. Identify faculty who will be pioneers in teaching innovation and who will do it well. There will be needs for special training and opportunities for faculty to gain experience. Faculty determine the content and format of the curriculum, but administration needs to persuade, facilitate, and reward innovative and effective ways of presenting the curriculum.

2. Administrators and faculty need to recognize that the student body is changing and that there is a need to define in the curriculum the amount of effort to be devoted to systems approaches. Also, graduate programs or other advanced technical programs traditionally build strength in specific disciplinary areas.

3. Professional societies need to be involved and can provide leadership. Although they are discipline oriented, the societies divide the infrastructure of individual institutions into a matrix to allow communication among faculty with common missions and perspectives. Societies also constitute a critical peer group beyond an individual campus.

4. Faculties must openly address the relationships among general education, professional education, and disciplinary specialization, especially with the goals of teaching students to think and interact in the process of lifelong learning.

5. Curricula and pedagogic approaches need to facilitate and effectively move more disciplines together so that they can become adopted by faculty and so that faculty can have a sense of ownership or belonging to a broader-based curriculum.

6. Faculty development is critical. There is a need to develop and support multidisciplinary efforts or retreats to gain faculty, student, and administrative perspectives on innovative approaches to the importance of teaching. Then, support mechanisms need to exist for experimentation and implementation of new methodologies.

7. Team teaching may help to develop the transition to systems approaches and solve short-range problems, but it adds complexity and does not address many of the real issues involved.

In a subtle way, the groups responded in a systems methodology through steps 1 and 2. Time limitations prevented comprehensive input and evaluation, however, and the groups were too large to have the proper discussion needed for systems evaluation. As in true student learning situations, some groups were dominated by strong individuals, and some individuals did not actively participate or offer input. Despite these limitations, the group reports contained a considerable overlap of outcomes, yet each report had a distinctive personality that reflected the makeup and background of the individuals who participated. One group, perhaps largely unknowingly, even treaded into developing a model that could be judged or evaluated, that is, step 3 of the systems approach.

In summary, the groups were able to use the initial steps of a systems approach in considering the use of a systems approach. The exercise helped the individuals to recognize the strengths and weaknesses of systems approaches and gave them a glimpse of how students may respond or interact in the analysis and synthesis settings. Above all, however, the presentation and discussion helped the audience gain a deeper appreciation for systems technologies and methodologies and how they can facilitate the teaching of agricultural science.

The Social and Ethical Context of Agriculture: Is It There and Can We Teach It?

Otto C. Doering III

James G. Leising, Rapporteur

"Values are the emotional rules by which a nation governs itself. Values summarize the accumulated folk wisdom by which a society organizes and disciplines itself. And values are the precious reminders that individuals obey to bring order and meaning into their personal lives. Without values, nations, societies and individuals can pitch straight to hell" (Michener, 1991:90–91). So argues James A. Michener in a plea for the teaching of family and community values. He sees such values as being distinct to each group in society and critical to the working of any society—with each group or family having something unique to convey to its next generation.

Many in agriculture still labor under the belief in agriculture's uniqueness. This notion rests on the recollection of Jefferson's concerns for a yeoman class to preserve democracy and on the continuation of agrarian beliefs from the era of populist fervor in America. The preservation of the myth of agricultural uniqueness also preserves the notion that agriculture has a unique social and ethical context. One aspect of this has been a belief that agriculture has a special relationship with the natural world. Depending on the values implicitly expressed, this relationship is variously described as one of harmony or one of management of nature, ranging from that of an English garden to that of Attila the Hun. Some other aspects involve the value of labor, the place of community, and the inherent value of rural life.

What we all need to face is the fact that the uniqueness is gone and that agriculture is unlikely to operate or exist on the basis of its own unique social and ethical contexts. Other social and ethical contexts have overtaken that of agriculture—if not by force of virtue, then by force of swamping the agricultural population with an industrial one. It was the social and ethical context of the industrial population that first swamped the agricultural population. The dynamism of the industrial context is superbly reflected by Carl Sandburg's poem "Chicago." Chicago is not only "Hog Butcher for the World" and "Stacker of Wheat" but also "Tool Maker . . . Player with Railroads . . . Stormy, husky, brawling, . . . Flinging magnetic curses amid the toil of piling job on job . . . Bareheaded,/Shoveling,/Wrecking,/Planning,/Building, breaking, rebuilding" (Sandburg, 1970). This is the new American industrial city. Yet this poem was written in 1916, only a decade after the Country Life Commission (Rasmussen, 1975), written at a time when half of the nation was still agrarian. We are now well beyond the industrial age and into a postindustrial age and context. To believe that agriculture still has its own unique social and ethical context is to be two revolutions out of date.

What we face now is not the challenge of dealing with a social and ethical context that flows from the agricultural experience. We are having to deal with the social and ethical context of someone else's age. Those whom we teach in schools of agriculture are not products of the farm—the Jeffersonian and populist vision—they are products of the postindustrial age who will be unlikely to go to the farm or anything like it.

Our current students come to us not only without a sense of agricultural context but also with only a very limited collection of any strong social and ethical norms. Neither families nor communities feel as strongly about inculcating a given set of norms in their young as was the case in more interdependent communities generations ago. As these "undernormed" students approach the universities or colleges of their choice, they are subject to the posture of modern science that tries to ignore or avoid dealing with social or ethical norms as much as possible. Science in schools of agriculture is not unique in its disinterest in social and ethical norms. Industrial science and scientific production also have no interest or willingness to deal with such norms unless they reinforce science or production.

Over the years, agriculture has had some touchstone social and ethical issues that have been barometers of our social and ethical sensibilities. Migrant labor has been one of these. The way we have dealt with this issue is by eliminating it—eliminating it without compensating the displaced, eliminating it without retraining the redundant, eliminating it because it was so socially embarrassing that we were unwilling to apply good management science to a first-class labor force and make them more productive. Church groups might give money to migrant labor organizing efforts, but the same diocese might not minister to the spiritual needs, let alone social needs, of the migrants. Few applied good science to make viable employment—instead we applied technological displacement. In spite of (or maybe because of) harping criticism from the outside, agriculture was unwilling to tackle this issue constructively. It seems ironic that having punted concerns about our treatment of people, we are now having to respond to concerns about battery of chickens and animal rights.

What would we like to have a student of agriculture be prepared to do about social and ethical questions? My goal would be to produce students who are able to recognize and weigh the trade-offs involved in applying broad social and ethical norms to today's decisions. I am concerned that our students are mostly unable to deal with issues like the "Big Green" referendum in California (banning a large number of agricultural chemicals and addressing many other environmental concerns), given the many conflicting social and ethical norms and goals involved. We cannot avoid responsibility by using the device of throwing our students into international or other cross cultural experiences to sensitize them to "real" issues—they must first be able to identify and deal with the context of their own society. Only later might a cross-cultural experience be instructive in further teaching the broad nature of our own norms. I learned more about American social and ethical norms during 2 years of postgraduate study at London University than at any other time in my life (but I started from the basis of a strong liberal arts undergraduate education).

Many of the faculty in schools of agriculture had a single response to the Big Green issue that was a non sequitur to the rest of society. The response was: "If the public were only adequately trained in science, they would recognize the need for these chemicals and the inherent safety of their use." This response is indicative of several tragic flaws in the way that the agricultural minority approaches the rest of society. First, even if a member of the general public had the same scientific knowledge as the scientist, that individual might still have a different risk preference and values with respect to health or concerns for environmental damage. Second, the general public is not likely to ever be adequately trained in science in the eyes of scientists. Third, we are dealing with public perceptions, which may or may not correspond to scientific facts and may correspond more closely to information from a source believed to be trustworthy. Being considered trustworthy is closer to a value judgment involving social and ethical norms than it is to scientific accuracy. Finally, my fellow scientists missed some of the major trade-offs and potential problems of Big Green by being concerned only about the scientific facts. Some of the major trade-offs were social and ethical, and neither extreme in the debate had all the angels on its side. By being unconcerned with the social and ethical arguments central to major public issues, we leave the game in forfeit.

One of the impacts of Big Green would have been to move the

production of a number of specialty vegetables and fruits to other states or countries. As a consumer, I know that California has the most stringent pesticide and fungicide use regulations in the nation. California also has the highest standards of safety and economic protection for its agricultural labor force of any state or country. If we move this production out of California, I am likely to be less well protected from excess chemicals or more risky chemicals than was the case when California grew the product. Certain foods might be more expensive or less available. The labor used to grow and harvest the product is also less likely to have good working conditions and adequate wages and benefits (according to my standards as an urban consumer) than would be the case in California. What this says is that, in some instances, Big Green would likely result in potentially more chemical exposure for the rest of the nation that now consumes produce from California and consumer choice would be changed by price and availability. It might also result in production under conditions much less favorable to agricultural workers—and those of us outside of agriculture know about these things: we boycotted grapes! All of these secondary impacts involve social and ethical considerations that are important to society.

How would our students look at Big Green? Would they temper the science-only approach? How arrogant might their science-only approach be? Would they recognize the importance of public perception and trust and know what they are based on? Would they be able to identify and assess countervailing social and ethical concerns even when one side appeared to have all the social and ethical weight in its favor? Could they make decisions on the basis of both science and nonscience?

As another example, how would our students deal with the "circle of poison" issue? Again, scientific and factual information appears to be on one side, which is pitted against social and ethical concerns on the other side, whose proponents suggest we stop the production of unregistered chemicals for export. However, even social and ethical norms can be of widely different scopes and contexts. There are ethical as well as good scientific arguments on both sides. Having worked in developing countries for a number of years, I do not feel I have the ethical right to tell a subsistence farmer that he is not allowed to use a chemical that is not registered with the U.S. Environmental Protection Agency when he is feeding a family suffering from malnutrition. I feel it would be arrogant of me to do so. Wealthy people are more able to worry about the long run than very poor people are. Do wealthy people have the ethical right to force the long-term view on the poor? I do not find this an easy question to answer. Those on both extremes in the debate are more comfortable than I am in easily dealing with or just ignoring such questions.

Can and should we teach an ethical and social context to our students? We can, but we probably should not. Can we teach a broad perspective that gives context to social and ethical issues? I am not sure that we can, but I think this is a better approach.

What I am saying is that, first, our society is providing less and less of an overall social and ethical context for our young. Few parents are willing to teach and convey it, and precollege schools believe that they are not allowed to teach it. Second, if we try to teach this at the college level, we might end up teaching a doctrine or set of personal professorial beliefs. However, I am less concerned about a Marxist in the classroom than I am about a bad Marxist in the classroom whose only appeal is to the heart, not the mind. Students need to be able to recognize and then analyze the social and ethical trade-offs inherent in any important decision to which they can then apply their own developing values and their knowledge of scientific facts. Winston Churchill's quip that "if a young student is not at first a Socialist he does not have a heart, but if he does not later become a Conservative he does not have a brain" tells us more than his view about socialists and conservatives. It says something about the process of learning and exploring values and balancing these with facts in decision making.

At the college level, we should not teach a given social and ethical context per se. We can broaden the knowledge base in which an individual deals with such questions and we can demonstrate an approach to such issues by example. Exposing undergraduates to teachers who are good in their disciplines, who have broad experience, and who have their own well-developed social and ethical context is one of the best learning experiences. What sorts of individuals fit the bill? Keith Kennedy, Jean McKelvey, Jean and Ken Robinson, Dan Sisler, and Milton Barnett at Cornell University; John Axtell, Bruce McKenzie, Deborah Brown, and Don Paarlberg at Purdue University; Emerson Babb and Bob Peart at the University of Florida; and Bill Chancellor and Sylvia Lane at the University of California at Davis have done this for earlier generations. If you want students to gain social and ethical context, do not try to teach it badly; put in the classroom teachers for whom it pervades the learning experience. Students then see what such a context does for one's ability to analyze difficult issues and make critical decisions. A course in ethics can broaden a student's scope but may offer little by example or experience in making difficult ethical choices.

On a curriculum level, there must also be some background information or personal experience giving the student some basis for making comparisons and choices. This means taking a good class in American government, some well-taught history, English that facilitates better reading and writing (allowing the student to enter the world of ideas), cultural anthropology, applied sociology, etc. Without something like this, a student has no context for social and ethical issues—no standards for comparison and discrimination other than personal emotions and experiences.

The following are some goals that we might set for ourselves to equip our students to integrate social and ethical factors into their decision making:

1. Do our students understand the difference between facts and values? Are they equally comfortable dealing with each, and do they recognize the role of each in decision making?

2. Do our students have a broad context for social and ethical questions in addition to their own personal beliefs, values, and experiences?

3. Are our students able to identify and assess trade-offs that involve facts and values, science, and the social and ethical context?

4. Have our students been sufficiently exposed to teachers who convey experience in dealing with social and ethical issues in a nonadvocacy, nonproselytizing way?

If we can answer yes to each of the above, we are turning out an individual ready to deal with the social and ethical context of the postindustrial world.

References

Michener, J. A. 1991. What is the secret of teaching values? Money (April): 90–91.

Rasmussen, W. 1975. Pp. 1860–1906 in Agriculture in the United States: A Documentary History, vol. 2. New York: Random House. (First published in Bailey, L., et al. 1909. Report of the Country Life Commission.)

Sandburg, C. 1970. Chicago. P. 3 in the Complete Poems of Carl Sandburg. New York: Harcourt Brace Jovanovich.

RAPPORTEUR'S SUMMARY

Many believe that agriculture has a unique social and ethical context. Some of the major aspects involve agriculture's special relationship with the natural world, the value of labor, the place of community, and the inherent value of rural life. Otto C. Doering III argued that the industrial segment of society has overtaken that of agriculture and has, in effect, replaced the social and ethical context for most Americans. He stated, "We are now well beyond the industrial age and into a postindustrial age and context. To believe that agriculture still has its own unique social and ethical context is to be two revolutions out of date."

Currently, students come to colleges of agriculture without a sense of agricultural context, but they also have limited collections of strong social and ethical norms. As a result, students are subject to the attitudes of modern science. These attitudes attempt to ignore or avoid the ideas of social and ethical norms and the roles they play in society. A good example of this attitude was the "Big Green" referendum in California that was aimed at limiting a large number of agricultural chemicals. Scientists responded from the perception that if the public were only adequately trained in science, they would recognize the need for these chemicals and the safety of their use. This type of logic often errs, because the scientific community fails to recognize that we are dealing with public perceptions that may or may not correspond to scientific facts and that may correspond more closely to information from sources believed to be trustworthy. Being trustworthy is closer to a value judgment involving social and ethical norms than it is to scientific accuracy.

What should students of agriculture be prepared to do about social and ethical questions? According to Doering, students should be able to recognize and weigh the trade-offs involved in applying social and ethical norms to today's decisions. He also advocated that pushing students into international or other cross-cultural experiences was not the answer. Rather, they must first identify and deal with the context of their own society.

Can and should we teach an ethical and social context to our students? Doering believes that we cannot teach the social and ethical context per se. Rather, he feels we can broaden the knowledge base in which an individual deals with such questions and we can demonstrate an approach to such issues by example. In other words, do not try to teach it badly, but instead, put in the classroom teachers who have it so that it pervades the learning experience. Students then see what such a context does for one's ability to analyze difficult issues and make critical decisions. It was concluded that teachers often attempt to separate fact from values rather than looking at all the information brought to bear on the issue that will cause students to look at the whole.

Much debate ensued in the discussion group over whether courses could be used to teach the values and ethics of agriculture. No clear consensus was evident, but it was pointed out that most faculty have little formal education in the area of ethics, and therefore, courses of this nature may need to be taught jointly with faculty from philosophy or other social science departments. A beginning point for implementing the teaching of ethics and values might be for each professor to agree to integrate topics that address the question, "What social and ethical implications does this course have for mankind and society at large?" However, this suggestion is made with the understanding that faculty would be given the opportunity for professional development in this area and would have thought deeply about the issues involved.

Another concern that was addressed was the idea that students do not have a context for dealing with social and ethical issues because they have no standards other than an emotional context for discriminating the various sides of an issue. In other words, should the curriculum include some general education in such disciplines as history and government, English, anthropology, and sociology to provide a context for considering issues? After much discussion, the consensus was yes. However, one question remained: "How many units should be required and what agricultural courses should be deleted to make room for general education courses?" It should be noted that most universities have required general education courses; however, this discussion was focused on the idea of creating a coherent core of courses specifically for agricultural students.

In summary, Doering suggested four goals that faculty members might set for students in the area of the social and ethical context of agriculture.

1. Students should understand the difference between facts and values.

2. Students need a broad context for social and ethical questions in addition to their own personal beliefs and values.

3. Students should be able to identify and assess trade-offs that involve facts and values, science, and the social and ethical context.

4. Students should be exposed to teachers who convey experience in dealing with social and ethical issues in a nonadvocacy, nonproselytizing way.

There is no question that the social and ethical context of agriculture is one of the least understood and least taught areas of the curriculum for agricultural students. This session provided a discussion of the issues involved and provided insight into teaching about the social and ethical context of agriculture. It is apparent, now more then ever before, that if we fail in dealing with these aspects of agriculture, we could become paralyzed as an industry.

The Economic Context
of Agriculture

James L. Rainey

Larry J. Connor, Rapporteur

Agriculture is often praised as a special way of life by farmers, some farm organizations, and others who take comfort in the tradition of American agriculture's homogeneous past.

Undeniably, there is much that is appealing about the freedom and rugged individualism that are at the center of agriculture's societal appeal. But the fact is that agriculture is also a way of making a living, and the imperatives of economics are a powerful force behind agriculture's rapidly changing competitive environment. Change is never easy, nor is it popular. There is resistance to change, even when economic forces would seem to dictate its inevitability.

Agriculture's leaders—producers, business managers, and certainly educators—must therefore accept their role and responsibility to be in the forefront of efforts to build understanding and acceptance of a rapid agricultural evolution. An evolution that is fueled by economies of scale and global competition. Consider what is happening to U.S. agriculture:

• The trend in consolidation continues, with fewer but larger farms and fewer but larger suppliers. It is estimated that by the end of this century, the United States will be provided with 60 to 70 percent of its agricultural products by less than 100,000 producers. At the same time, there will be more hobby farms. The traditional, diversified midsize farm unit is disappearing.

• Rural transition continues to occur as small communities shrink and regional market centers grow.

• Agricultural policies are shifting to a greater emphasis on market orientation and lower subsidiaries, with greater regulations because of environmental and food safety concerns.

245

• Technology shifts are occurring in the forms of computer and electronic transfer capabilities and larger, more efficient production and processing equipment.

• Biotechnology is changing what farmers grow, how they feed and protect what they grow, for whom they grow, and the ultimate end uses of their products.

• Agribusiness consolidation will continue, with a dominance of a limited number of large food system companies. If present trends continue, the top 50 firms will own 90 percent or more of food industry assets by the end of this decade.

• Coordinated supply and marketing will continue to expand. With decreasing supply and marketing options, production contracts and food system integration will increase.

• Revenues will be concentrated in the "farm gate-to-consumer" sector. While the consumer's total market basket cost increased 65 percent from 1972 to 1988 (in 1972 constant dollars) and while the cost of food as a percentage of disposable income has continued to go down, farmers are receiving 4 percent less in total market basket revenues.

With this kind of evolutionary, almost revolutionary, change taking place, it is easy to understand how important our task is to find—and empower through education—the kind of people who can make the difference in the demanding agricultural arena.

The sad fact is that business is finding it tougher and tougher to identify, retain, and reward people who are equipped to compete in a technology-driven world economy. The agribusiness industry can be hit especially hard. The U.S. Department of Agriculture projects an annual shortage of 4,000 college graduates for some 48,000 agriculture-related jobs into the 1990s. Many companies that have cut back over the last decade are starting to hire again. The problem is that they are simply not finding enough people. We need to get serious on why this is so and what we should do about it. We must persuade more young people, from the elementary level through college, to catch the vision and understand that we need their brainpower to manage change, solve problems, make decisions, analyze data, and create new products.

Statistics indicate that, in the main, schools are coming up short. The reason may come easy for some of us, perhaps too easy. We can attribute it to the conditions in inner cities that put many minority children at a disadvantage, poorly funded schools, and underpaid and, in some cases, underqualified teachers. Whatever the reason, the fact remains that 1 million students quit school each year, and of the 2.5 million students who get through, one-fourth cannot read beyond the eighth-grade level.

Industry does have a role and an obligation to participate in the business of education by adopting schools, providing internships, and letting the schools know what types of skills workers will need. Some businesses are already pitching in.

With that in mind, I offer eight points that I think are crucial to ensuring an adequate supply of people to provide leadership for agriculture in the years ahead. Perhaps these points will serve to provoke our thinking about the academic–business partnership that is needed to meet future educational challenges.

First, we have overemphasized the importance of academic performance as the key determinant of a student's potential for development in the business environment. It is a fact that many good students do not progress in the business community. Quite often, there are two reasons: a lack of communications skills and an inability to get along with people. How do we emphasize the need for well-rounded people who are equally adept academically and on an interpersonal relationship level?

In people-intensive businesses, the most important management challenge is people. If these otherwise gifted people cannot communicate effectively or get along with their peers, great potential resources can be lost. Such problems might be eased if these bright and skilled people were given greater exposure to such courses in college as basic psychology, communications, and team building.

Second, we need to do a better job of identifying people with managerial and technical developmental potential so that we can move them through a broader spectrum of experience in a shorter time and thus reward them more quickly—and retain them in the process.

Although job skills are becoming increasingly specialized and the natural inclination is for employees to become comfortable with their specialty, the need for broadening the skills of managers in an expanding menu of business disciplines is apparent. Perhaps business and academia can work together more effectively to identify, test, and confirm the developmental potential of key individuals and thus go on to expose them to specialized advanced training.

Third, we need to identify more quickly the reasons why so many young Americans lose interest in pursuing careers that are so critical to the strength of our economy and our global effectiveness. Business and academia need to team up to learn how to identify promising people at an early point in their education and development and thereby help to steer them into rewarding and successful careers. It seems to me that it is feasible for industry and academia to find long-term programs to educate our youth, beginning in elementary schools, with respect to the opportunities that may be available if they pursue specifically identified training in college.

Fourth, we need to acquaint undergraduate and graduate students with the reality of "combat duty" in the business community. We need a broader variety of internship programs to equip them for frontline hazards.

247

Fifth, there is a desperate need to get youth in the United States more interested in what a global "one-world" economy means and the significance of international trade. Despite the excitement that such a career could hold, the percentage of graduates from either high school or college who have some foreign language capability is very small. Why? How do we make available to our students greater opportunities for study and work experience abroad?

Another challenge that merits our mutual concern is the fact that so many young Americans lose interest in subjects that are so crucial to the U.S. economy and our global effectiveness. I refer to the fact that we are graduating a declining number of engineers in comparison with other countries. And at a time when our food and agricultural system is open to great opportunity, enrollment in our schools of agriculture has declined steadily over the last 5 years. If the agricultural curriculum is to keep pace with the employment demands of the agricultural and food sector, it also needs to produce graduates familiar with economics, business market analysis, sales and advertising, computer science, and business management.

Points six and seven are mirror images. What can academia and industry do to improve executive education, both development and training. We have agreed, I think, that we must mutually support efforts to attract students, keep them in school, and prepare them well for their business careers. Is there not also a need and opportunity for continuing executive education and for advanced training in business disciplines for corporate board members?

Conversely, does industry have a role to play in improving the quality of educators? Is it too far a stretch to suggest that professors might benefit from an introduction to the real-world business environment?

Finally, industry and academia must more effectively finance and commit to basic research and development in areas that will best serve our industrial and consumer needs. Industrial spending on research and development rose only 1 percent in 1990, totaling $74 billion. We need these new minds to help us to get the most from our spending on research and development. More of it could be dedicated to that rapidly growing field known as "speed to market"—the reorganization of companies to accelerate the rate at which they can produce new goods and services. Increasingly, profits are coming from new products.

I close with an observation by human relations expert Ralph Stayer in the *Harvard Business Review*. "People want to be great," he says. "If they aren't, it's because management won't let them be" (Stayer, 1990:82). I think enlightened management today is quite receptive to the fact that people want to be great. Companies that attract the right people reward them not with just money, but they empower them or grant them the directed autonomy to be great.

248

Tragic but true, the millions of unmotivated American youth represent a great waste to American agriculture. We must find better ways to tap and unlock their potential to deal with the forces of change that are reshaping the economic context of agriculture.

Reference

Stayer, R. 1990. How I learned to let my workers lead. Harvard Business Review (Nov-Dec):66–83.

RAPPORTEUR'S SUMMARY

James L. Rainey initiated the session by emphasizing the importance that leaders in agriculture accept their roles and responsibilities in building an understanding and acceptance of a rapid agricultural evolution that is fueled by scale economies and global competition. He outlined developments that will have an impact on U.S. agriculture and emphasized the importance of industry in having a role and being obliged to participate in the business of education by adopting schools, providing internships, and informing schools as to the type of skills workers need in the marketplace. He emphasized eight points that are crucial in ensuring an adequate supply of people to provide leadership for agriculture in the years ahead.

The subsequent discussion centered around students who enter the world of work in business and included the following items:

• How does one get industry and academia to link and form relationships? Various alternatives were discussed, particularly the use of sabbaticals. It was agreed by all parties that a greater commitment is needed by industry.

• Business firms need to keep a presence on campus, even when they are not hiring. During the agricultural recession in the early 1980s, many firms ceased recruiting efforts. It subsequently became difficult for them to resume their recruiting activities.

• Agricultural graduates do a good job in entry-level positions, but they often have difficulties in competing for upper-level management or chief executive officer positions. It was agreed that there is a critical need to identify students on campus and in business firms much earlier and help them move through a variety of disciplinary experiences in preparation for upper-level management positions. However, stiffer competition can be expected as individuals move up through the various managerial levels.

• How do students obtain the necessary "combat" experience in business firms? Various options were discussed, including the use of mandatory and optional internships, the expansion of a college education beyond 4 years so that students may have business experience prior to graduation, or the use of cooperative education agreements between colleges and business firms (examples would be the programs at Rutgers University and the University of California at Davis). It was also agreed that faculty can contribute by becoming more knowledgeable by spending some time in companies and living in the "trenches."

• There was a consensus that typical shortcomings are communications skills in graduating agricultural students who enter the world of business. Suggested communications skills priorities were listening (what is not said as well as what is said), writing, and speaking. The importance of faculty of colleges of agriculture in dealing with English and communications skills, and not just leaving these to the communications faculty, was strongly emphasized.

• The last discussion item dealt with interpersonal skills for students who enter the world of business. How do students get along with other people without becoming "yes" people? The importance of college courses dealing with basic psychology, teamwork, and communications was again stressed. It was also suggested that students need to learn the "sphere of influence" needed to get a job done and work with people. Students must also learn that working effectively with others is a critical element of their accountability and success.

The Global Context
of Agriculture

Edna L. McBreen

Susan G. Schram, Rapporteur

The internationalization of U.S. universities has become a nation-wide cause célèbre, and not a minute too soon.

As the country's concern with economic competitiveness increases with every American purchase of a Japanese vehicle, government and business leaders are increasingly critical of the U.S. educational system and the narrow isolationist knowledge base of our graduates, who are our primary product.

The actions leading up to the Persian Gulf War, the military successes of that war compared with a relative lack of political and diplomatic successes, and the general lack of U.S. understanding of the social, cultural, and political aspects of the Middle East all lead us to conclude that our narrow perspective of the world will limit our political and diplomatic successes.

The opening up of Eastern Europe and the enthusiastic anticipation of growing consumer markets, teamed with the personal, heartfelt concern of Americans who trace their ancestry to Eastern Europe for the economic and political trials currently being experienced in that part of the world, have increased interest in area studies and languages related to Eastern Europe.

Even though internationalization is highly acceptable in the higher education community, the definition and implementation of the internationalization process are still a long way from being complete. In their report "Internationalizing U.S. Universities—Preliminary Summary of a National Study," Henson and coauthors (1990) reported on a major survey of U.S. universities.

As one of the first studies of its kind, the results are by no means conclusive. They do, however, give us some much-needed insight into the internationalization process and the support that exists for that process. For example, in reporting priorities for international activities by upper administration officials (deans and those in higher positions), of the 13 priorities listed, only 4 were directly related to the undergraduate curriculum: (1) encouragement of foreign language study, (2) inclusion of international content and materials in the curriculum, (3) offering study or internship abroad opportunities for U.S. students, and (4) establishing or implementing area studies programs.

The other priorities, such as recruitment and training of international students, support of faculty exchanges, and participation in development assistance projects, potentially could have an impact on the undergraduate curriculum; but they have not always done so in the past.

Even though foreign language, inclusion of international content in the curriculum, and study abroad were selected among the top 4 of the 13 priorities, progress in these areas appears to be somewhat inconsistent. While universities are experiencing significant increases in enrollment in foreign language programs and some increases may be occurring in study abroad programs (Zikopoulos, 1990), the report of Henson and colleagues (1990) was not terribly enthusiastic about the extent of the internationalization process in the undergraduate curriculum. In reporting the subjective results of the survey and follow-up interviews, some of the glitches in the process became apparent; for example, the incorporation of international content into the undergraduate curriculum does not appear to have progressed very far at many institutions, and most respondents have not determined or defined what the optimum level of international competence for graduates should be, while even fewer are beginning to evaluate that competence. It is, perhaps, this lack of understanding of the optimum levels of international competence for graduates that is the major stumbling block for internationalization of the undergraduate curriculum—and it may be the most important stumbling block to internationalizing the undergraduate curriculum in agriculture.

Agriculture itself has become more and more subject to international trends and issues over the last decade, with predicted increases in that direction. In the December 27, 1990, *Kiplinger Agriculture Letter* (The Kiplinger Washington Editors, Inc., 1990), which focused on agricultural research, developments were described in glowing terms, but news related to the international competitiveness of U.S. agricultural research was not so encouraging, as illustrated by the facts that the number of U.S. students who are training to be agricultural scientists has dwindled, and U.S. investment in agricultural research as a percentage of nondefense research expenditures lags behind that of the European Community, Canada, and Australia.

It has, no doubt, never been really appropriate to assume that a monopoly of agricultural expertise exists in the United States—it is even less appropriate now.

The U.S. agricultural industry has long been involved in export trade. In fiscal year 1989–1990, exports of 10 major commodities amounted to over $21 billion (Food and Agricultural Policy Research Institute, 1991). The volume of agricultural exports is projected to rise steadily after 1990–1991 because of the projected growth in international income and population.

U.S. agriculture has always been subject to the realities of a global environment and the lack of respect that various agricultural pests have for political boundaries. Elementary school children in Texas understand that point when they sing a song about the boll weevil "looking for a home" and crossing the border into Texas.

There appears to be no lack of understanding of the international trade impact on agriculture, there is considerable recognition of global environmental impacts, and there is a growing awareness of the value for U.S. agriculture of the international agricultural research community's work. Yet, almost no real progress has been made to ensure that graduates of our colleges of agriculture are internationally literate.

On the other hand, many colleges of agriculture are beginning to consider the problem and the development of solutions to that problem. We acknowledge, for instance, that students in agriculture make up less than 1 percent of U.S. students who study abroad. Some colleges of agriculture are embarking on the development of international opportunities specially planned for agricultural students.

There are still many questions to be answered in planning the internationalization of undergraduate education in our colleges of agriculture—but, surely, there are no questions about the need to do so. Some of the following are planning issues that need to be addressed:

• With an ongoing commitment to serve U.S. agriculture and the respective states, what should be the overall level of commitment of U.S. colleges of agriculture to an understanding of international agriculture?

• What are the levels of understanding and competence in international agriculture that every graduate of a college of agriculture in the United States should attain?

• What are the elements of international agriculture that should be incorporated into the undergraduate curriculum?

• What role should study abroad and other international opportunities play in the undergraduate agricultural curriculum? How can those opportunities be specially tailored to the needs and career goals of agricultural students?

• To what extent should colleges of agriculture rely on other departments and colleges in the university for internationally oriented courses?

• Do different majors within agriculture have different needs for internationalization? What are the international competencies that graduates in the various majors should attain?

References

Food and Agricultural Policy Research Institute. 1991. FAPRI 1991 U.S. Agricultural Outlook. Ames: Iowa State University, and Columbia: University of Missouri-Columbia.

Henson, J. B., J. C. Noel, T. E. Gillard-Byers, and M. I. Ingle. 1990. Internationalizing U.S. universities—Preliminary summary, of a national study. In Internationalizing U.S. Universities: A Time for Leadership. Conference Proceedings, June 5–7, 1991, Spokane, Wash. Spokane: University of Washington.

The Kiplinger Washington Editors, Inc. 1990. Kiplinger Agriculture Letter 61(26):1–2.

Zikopoulos, M., ed. 1990. Open Doors: 1988–1989. New York: Institute of International Education.

RAPPORTEUR'S SUMMARY

In the current era of global competition and interdependence, there is growing awareness that the United States requires a higher level of international competence. The educational system at all levels is instrumental in raising those standards of international competence.

The discussion in the Global Context of Agriculture group, led by provocateur Edna L. McBreen, centered around the broad challenge of internationalizing the university, and specifically, the agricultural curriculum, toward the goal of adequately preparing agricultural students for competing well in the global economy.

As noted by McBreen, although we are making progress in internationalizing the curriculum, there is still a shortage of U.S. students trained to deal with the international context. Exposure to the international context is particularly inadequate at the undergraduate level. In agriculture, for example, only a small percentage of students participate in study abroad programs.

Internationalization of the Curriculum

The group agreed that attempts to internationalize the agricultural curriculum currently take many forms. It may be requiring an international experience abroad as part of undergraduate study; it may be placing an international emphasis on a conventional major (toward the goal, for example, of preparing an agronomist to graduate with a good set of international "tools"); it may simply be teaching more comparatively—for example, how the soils of Mississippi compare with soils around the world. Some contend that internationalization is merely the inclusion of a foreign language in the undergraduate curriculum requirements, although that was not viewed as adequate by the discussion group.

It was clear that agriculture seeks to marry the *intellectual* with the *practical*, including the following:

- academic course work combined with application through field trips abroad;
- language training, not for its own sake, but for preparation for pending international field experience; and
- semester abroad programs in which academic course work is combined with study of a different culture.

Several examples of low-cost collaborative efforts on the campus also exist. At one university, for example, several colleges pool energies to bring resource people to the school for seminars. In some cases, students and faculty study abroad together. One state sponsored joint international seminars for both students and faculty; another sponsored Fulbright cross-discussion group projects.

One group participant summarized three actions that were needed to further internationalize the agriculture curriculum: (1) emphasize international interconnectedness in all agricultural courses, (2) help undergraduates to appreciate other cultures, and (3) develop new international courses. It was noted that the first two of these goals could be accomplished without additional resources.

Internationalization of Faculty

Internationalizing the faculty is key to internationalizing the curriculum. In some states, sophisticated agricultural producers are more internationally involved and aware than our faculty. Perhaps faculty need to interact to a greater extent with industry representatives to assess industry's needs and discover how to better prepare globally competent students.

The importance of the international perspective should be reflected in tenure policy and should be stated in the missions of colleges of agriculture. Some states are very progressive in this regard. One state legislature has taken steps to encourage that rewards be given to faculty with international experience.

Faculty exchange is important, not only for the personal enrichment of the faculty but also for the potential contribution to the internationalization of the university. One participant suggested

255

that the international competencies sought in students needed to be defined, and then a process for internationalizing faculty should be developed.

Conclusion

In summary, the group discussion indicated that there are many encouraging signs emanating from our colleges of agriculture and innovative steps being taken to internationalize the curriculum. In many ways, agriculture is avoiding the mistakes of others. For example, study abroad is not viewed merely as a vacation. In addition, agriculturalists are asking, "internationalization for what?" The college of agriculture approach is practical. We look to the result: How will the degree ultimately be used? How will international competence enhance students' employability in various fields? We have recognized that only one component, that is, 2 years of a foreign language, will not be totally adequate.

Innovations in internationalization have employed flexibility and collaboration—collaboration with colleges other than colleges of agriculture and with other institutions to capitalize on their strengths. Perhaps further collaboration is needed between graduate and undergraduate programs and through technological innovation such as AG*SAT (Agricultural Satellite Corporation, which links via satellite 34 U.S. land-grant universities so that they can share academic instruction, cooperative extension programming, and agricultural research information).

The following challenges remain to be addressed:

• How do all of the various international activities on campus fit together toward the internationalization of a campus?

• How does the long tradition of work in international development through Title XII and other programs enter into the equation? How can it enrich the academic curriculum?

• How does the university capitalize on the experience of faculty members returning from a development experience?

• How do Title VI area studies programs relate to the total picture?

• How will the U.S. Agency for International Development's new Center for University Collaboration in Development help to internationalize a campus and stimulate linkages between the various departments on a campus?

• Is there a need for a clearinghouse of ideas on how to internationalize the agricultural curriculum?

Appendixes

A

Program Participants

C. EUGENE ALLEN joined the University of Minnesota faculty in 1967 and became dean of the College of Agriculture in 1985. He is currently vice-president for the Institute of Agriculture, Forestry, and Home Economics and director of the Minnesota Agricultural Experiment Station. Allen has also been a faculty member at the University of Minnesota in two departments—animal science and food science and nutrition. His research on animal growth biology and the functional and nutritional characteristics of animal food products is internationally known. Allen's research has been recognized through numerous awards from professional societies and has been recognized by a Distinguished Teacher Award from the College of Agriculture and the all-university Morse-Amoco Award for Outstanding Contributions to Undergraduate Education. His national leadership activities include major offices or initiatives for the National Academy of Sciences related to animal growth biology, agricultural science policy, food technology, and the role of food and agriculture. He presently serves on the Board on Agriculture of the National Research Council, National Academy of Sciences. Allen received a bachelor's degree from the University of Idaho and M.S. and Ph.D. degrees from the University of Wisconsin.

ROY G. ARNOLD is provost and vice-president for academic affairs at Oregon State University. He was formerly dean of the College of Agricultural Sciences at Oregon State University. He began this appointment on December 1, 1987, following 20 years as a faculty member and administrator at the University of Nebraska-Lincoln. His assignments at Nebraska included teaching, research, extension, head of the Department of Food Science and Technology, dean and director of the Agricultural Experiment Station, and vice-chancellor of the Institute of Agriculture and Natural Resources. Arnold received a B.S. degree from the University of Nebraska and M.S. and Ph.D. degrees from Oregon State University. He was presented with awards for outstanding teaching by the University of Nebraska and the Institute of Food Technologists. In Nebraska, Arnold worked closely with the Department of Economic Development and Agriculture to establish a center for food processing, marketing, and transportation. He has also interacted with various commodity and trade organizations in planning economic development initiatives.

KARL G. BRANDT is associate dean of agriculture and director of academic programs at Purdue University, where he also holds the rank of professor of biochemistry. He earned a B.A. degree in chemistry at Rice University in 1960 and a Ph.D. degree at the Massachusetts Institute of Technology in 1964, majoring in organic chemistry with a biochemistry minor. Following postdoctoral work at Cornell University, Brandt joined the biochemistry faculty at Purdue University in 1966 and was promoted to the rank of professor in 1975. From 1981 to 1984 he served as assistant dean of the graduate school. He accepted his current position in 1984. In addition to his administrative duties, Brandt teaches an undergraduate biochemistry course each year. He has been recognized for excellence in teaching and counseling. His research expertise is in the area of kinetics of enzyme-catalyzed oxidation-reduction reactions.

WILLIAM P. BROWNE was visiting scholar at the National Center for Food and Agricultural Policy of Resources for the Future in Washington, D.C., through December 1991. He has resumed his duties as professor of political science and director of the master of public administration program at Central Michigan University (CMU). Browne began his association with CMU as an assistant professor in 1971. Since 1973 he has also advised private-sector and governmental units and agricultural sciences societies and foundations in areas such as citizen participation, grants, staffing, executive development, and agriculture and rural policy analysis. From 1985 to 1986, Browne was a visiting scholar with the Farm and Rural Economy Branch of the Economic Research Service, U.S. Department of Agriculture, where he coordinated an interest groups project concerning the Food Security Act. He earned his M.S. degree in political science in 1969 from Iowa State University and his Ph.D. degree from Washington University, St. Louis, Missouri, in 1971. In 1989 he received an Outstanding Academic Book Award for *Private Interests, Public Policy and American Agriculture* (Lawrence: University Press of Kansas, 1988). He has received 11 Creative Endeavor Awards at CMU, most recently in 1990, for education policy research, interest group politics, and public administration education.

BRIAN F. CHABOT is director for research at the College of Agriculture and Life Sciences and director of the Cornell University Agricultural Experiment Station. He is a professor in the Division of Biological Sciences at Cornell University. He received his degrees from the College of William and Mary and Duke University. His research focus was on the ecology of native and agricultural plants, with extensive work on the impact of the environment on plant growth and physiology. Chabot played a leading role in establishing the Low-Input Sustainable Agriculture Program (LISA) of the U.S. Department of Agriculture. He currently serves on the advisory committee for the Northeast LISA Program and on several national committees dealing with sustainable agriculture.

LYNNE V. CHENEY is serving her second 4-year term as chairman of the National Endowment for the Humanities (NEH). Her first term began in May 1986. Cheney directs the independent federal agency that provides grants to scholars, colleges, museums, libraries, and other cultural institutions to support research, education, preservation, and public programs in the humanities. As NEH chairman, Cheney has written four major reports, including *Tyrannical Machines* (Washington, D.C.: National Endowment for the Humanities, 1990), which assesses current problems in American schools, colleges, and universities and describes various promising efforts to institute reforms, and *50 Hours: A Core Curriculum for College Students* (Washington, D.C.: National Endowment for the Humanities, 1989), which urges U.S. institutions of higher education to strengthen course requirements so that undergraduates study essential areas of knowledge. Under her leadership, the NEH has launched several programs aimed at improving education in America's schools, colleges, and universities. Cheney has written and spoken often about American education and the value of the humanities to one's professional and personal life. Before coming to NEH, Cheney taught at several colleges and universities and was a magazine editor and widely published author. She earned a bachelor's degree from Colorado College and a master's degree from the University of Colorado. She received a doctoral degree, with a specialization in nineteenth-century British literature, from the University of Wisconsin in 1970.

JERRY A. CHERRY is professor and head of the Department of Poultry Science and chairman of the Division of Poultry Science at the University of Georgia. Born in Dayton, Texas, he received B.S. and Ph.D. degrees from Sam Houston State College and the University of Missouri in 1964 and 1972, respectively. In 1972, he joined the faculty of the Department of Poultry Science at Virginia Polytechnic Institute and State University as assistant professor. He was named associate professor and professor in 1978 and 1984, respectively. Active in both undergraduate and graduate education, Cherry received a certificate of teaching excellence from the Virginia Tech Academy of Teaching Excellence.

261

LARRY J. CONNOR is dean for resident instruction and dean of the College of Agriculture of the Institute of Food and Agricultural Sciences, University of Florida, Gainesville. He was formerly professor of agricultural economics and assistant director for planning, Agricultural Experiment Station, at Michigan State University. As assistant director for planning, Connor specified and developed priority research areas, expert teams, project proposals, and funding. His own research interests are in the areas of agricultural production economics, farm management, and agricultural resource economics. Connor's committee assignments at Michigan State have included chairing the Task Force for Curricular Revitalization, College of Agriculture and Natural Resources, and chairing the Admissions Policy and Entrance Requirements Task Force. He has also served on numerous national committees. Prior to joining the Michigan State University faculty in 1966, Connor was an agricultural economist with the Economic Research Service, U.S. Department of Agriculture. He earned a B.S. degree from the University of Nebraska and M.S. and Ph.D. degrees from Oklahoma State University.

OTTO C. DOERING III is professor of agricultural economics at Purdue University, where he teaches at both the graduate and undergraduate levels and conducts research on policy issues related to energy and resource use. His international experience is in food and resource policy. In 1990, as visiting scholar with the Resources and Technology Division of the Economic Research Service, U.S. Department of Agriculture (USDA), Doering assisted with analysis for resource and environmental issues in the 1990 Farm Bill. He was also on leave from Purdue in 1981 and 1982 when he was visiting scholar at the University of California, Berkeley, studying trade and resource issues. Doering was awarded the Distinguished Policy Contribution Award by the American Agricultural Economics Association in both 1978 and 1990, their Extension Economics Teaching Award in 1977, and recognitions for quality communication in 1979 and 1981. Doering is past director of the American Agricultural Economics Association, has served on national advisory boards for USDA and the U.S. Department of Energy, and has been a consultant to the National Academy of Sciences and the congressional Office of Technology Assessment. He is a member of Cornell University's College of Arts and Science Advisory Council and has served as chairman of the National Public Policy Education Committee. His academic training includes a B.A. degree from Cornell University, an M.S. degree from the London School of Economics, and a Ph.D. degree in agricultural economics from Cornell University.

FRANCILLE M. FIREBAUGH became dean of the College of Human Ecology at Cornell University in 1988; she is the seventh person to hold that title in the 65-year history of the college. Firebaugh received a Ph.D. degree from Cornell in 1962. She returned to Cornell to become dean after 26 years at Ohio State University, where she gained a reputation as a seasoned administrator. While at Ohio she served most recently as vice-provost for international affairs and was honored with a Distinguished Service Award at summer commencement in 1990. Firebaugh's academic life has a strong global flavor. Teaching and consultancies have taken her to Afghanistan, Egypt, India, and Malaysia. In 1988, Phi Beta Delta, an honor society of international affairs, awarded Firebaugh its first Faculty Award for Outstanding Accomplishments. She is a specialist in the area of family resource management, the author or coauthor of many scholarly articles, and coauthor of two books.

ROBERT M. GOODMAN is scholar-in-residence at the National Research Council and a visiting professor at the University of Wisconsin-Madison. He is a member of the Board on Agriculture of the National Research Council as well as the board of directors of the Cornell Research Foundation, Inc., and of Genetic Resources Communications Systems, Inc. From 1982 to 1990, he was vice-president and then executive vice-president with responsibilities for research and development at Calgene, Inc. Previously, he was on the faculty of the University of Illinois at Urbana-Champaign, where he was assistant professor (1974 to 1978), associate professor (1978 to 1981), and professor of plant pathology and a staff member of the International Soybean Program. He did undergraduate work at Johns Hopkins and Cornell universities. He received a Ph.D. degree from Cornell in 1973 and was a postdoctoral fellow in the Department of Virus Research at the John Innes Institute in the United Kingdom. Goodman's research has dealt with several aspects of plant virology and disease resistance. His work on the cause of bean golden mosaic disease led to his discovery of a new plant virus family containing single-stranded DNA genomes. The single-stranded DNA viruses of plants, called the geminiviruses, are now recognized as having major agricultural importance and as being an important tool in plant molecular biology.

JOHN C. GORDON is dean and professor of forestry and environmental studies at the Yale University School of Forestry and Environmental Studies. Gordon began his professional career as a plant physiologist with the U.S. Department of Agriculture Forest Service, North Central Forest Experiment Station. In 1970, he joined the faculty at Iowa State University, where he advanced from associate professor to professor, and then moved on to Oregon State University, where he was department head and professor of forest science. He earned a B.S. degree in forestry in 1961 and a Ph.D. degree in plant physiology in 1966, both from Iowa State University. Gordon's research is documented in over 90 papers, chapters,

263

and books, and he frequently lectures on research topics within the general area of tree physiology. He is currently leading and directly participating in research on biological productivity in the Copper River Delta in Alaska, with the cooperation of the Pacific Northwest Forestry Research Station. He teaches courses in research methods, agroforestry, and leadership. Consulting activities include business, government, and private, nonprofit organizations.

JO HANDELSMAN is assistant professor of plant pathology at the University of Wisconsin-Madison. Previously, she did postdoctoral work in the Department of Plant Pathology at Madison, which was supported by fellowships from the National Institutes of Health and the American Cancer Society. Handelsman's undergraduate work was in agronomy at Cornell University, and in 1984 she earned a Ph.D. degree in molecular biology from the University of Wisconsin-Madison. Her research group studies the molecular and genetic basis of microbe interactions with plants. Handelsman teaches a graduate course in phytobacteriology in which she uses the current literature of plant pathology to teach students, through analysis and discovery, how plant pathology draws on and contributes to the broader principles of genetics, biochemistry, ecology, and systematics. She also teaches an undergraduate course entitled Plants, Parasites, and People that uses examples from plant pathology, both historical and contemporary, to explore the social context of the uses of technology in agriculture. Handelsman is also a faculty participant in the development of a new certificate program entitled Agriculture, Technology, and Society.

NILS HASSELMO is the thirteenth president of the University of Minnesota. Born in 1931 in Köla, Sweden, Hasselmo was introduced to U.S. culture through the novels of James Fenimore Cooper and Mark Twain. As a student of Scandinavian languages and literature at Uppsala University, Hasselmo received the Mauritzon Fellowship for study at Augustana College in Rock Island, Illinois, a college founded in 1860 by Swedish immigrants. There he received a B.A. degree in 1957 and then returned to the United States on a fellowship from Harvard University to earn a Ph.D. in linguistics in 1961. In 1965 he came to the University of Minnesota as an associate professor of Scandinavian languages and literature. During 18 years at Minnesota, Hasselmo served as chair of the Scandinavian Department, associate dean of the College of Liberal Arts, and vice-president for administration and planning. In 1983 Hasselmo left Minnesota to become senior vice-president for academic affairs and provost at the University of Arizona, where he was known as a skillful negotiator and innovative policymaker. He returned to the University of Minnesota as its president in December 1988.

ROBERT M. HAZEN is a research scientist at the Carnegie Institution's Geophysical Laboratory in Washington, D.C., and professor of earth science at George Mason University. He received B.S. and S.M. degrees in geology at the Massachusetts Institute of Technology and a Ph.D. degree in earth science at Harvard University. After a year of studies as a North Atlantic Treaty Organization postdoctoral fellow at Cambridge University in England, he joined the Carnegie Institution's mineral physics research effort. Hazen is author of more than 160 articles and 7 books on earth science, materials science, history, and music. His research focuses on the close relationship between atomic structure and physical properties of materials. He recently led the Carnegie Institution team that discovered the identities of several record-breaking, high-temperature superconductors. Hazen's books have received widespread critical praise, and he is active in presenting science to a general audience. At George Mason University he has worked closely with James Trefil in developing a course on scientific literacy and a companion text, *Science Matters: Achieving Scientific Literacy* (New York: Doubleday, 1991). He teaches a course on symmetry in art and science for undergraduates and developed a methods course for public school science teachers in the District of Columbia. Hazen serves on the board of advisers for the National Academy of Science's National Science Resources Center and is a writer of the National Science Foundation's class materials that are distributed during Science and Technology Week.

RICHARD A. HERRETT is a private consultant in the areas of agriculture and the environment. He was formerly government relations scientific liaison for the Government Relations Office of ICI Americas, Inc., in Washington, D.C. In this position, which he assumed on February 1, 1987, Herrett was responsible for representing the company's technical interests on a range of issues, including agriculture, bioscience, and the environment before the appropriate regulatory and legislative bodies. Herrett joined ICI in 1970 as technical manager for the Agricultural Chemical Division in Goldsboro, North Carolina. In 1975 he assumed the position of director of research and development for the Agricultural Chemicals Division. A 1954 graduate of Rutgers University, Herrett holds a master's degree in agronomy and a Ph.D. degree in plant biochemistry from the University of Minnesota. He is currently chairman of the Chemical Manufacturer's Association (CMA) Task Force on Global Climate Change and of the Association of Biotechnology Companies' Agricultural and Environment Committee. Herrett is also president of the C. V. Riley Memorial Foundation, a nonprofit organization that promotes dialogue about major agricultural policy issues.

CHARLES E. HESS was sworn in as assistant secretary for science and education on May 22, 1989. He is responsible for the U.S. Department of Agriculture's research and education programs in the food and agricultural sciences, including general supervision of the Agricultural Research Service, the Cooperative State Research Service, the Extension Service, and the National Agricultural Library. Hess began his career with the Department of Horticulture at Purdue University in 1958. In 1966 he moved to Rutgers University, where he served as both associate dean and acting dean of the College of Agriculture and Environmental Sciences before becoming the first dean of Cook College at Rutgers. From 1971 to 1975, he was also director of the New Jersey Agricultural Experiment Station. In 1975 Hess was appointed dean of the College of Agricultural and Environmental Sciences at the University of California at Davis and associate director of the California Agricultural Experiment Station. In 1988, he assumed the additional post of director of programs, Division of Agriculture and Natural Resources, of the California Agricultural Experiment Station and the Cooperative Extension Service. Hess has served on the National Science Board of the National Science Foundation and as cochairman of the Joint Council on Food and Agricultural Sciences. He also chaired the National Research Council's Committee on a National Strategy for Biotechnology in Agriculture.

WILLIAM P. HYTCHE is chancellor of the University of Maryland, Eastern Shore (UMES), a position he was appointed to in June 1976 after serving as acting chancellor since 1975. Hytche received a B.S. degree from Langston University and M.S. and Ph.D. degrees from Oklahoma State University. He also studied at the University of Heidelberg, Oklahoma University, Oberlin College, and the University of Wisconsin-Madison. Hytche came to UMES, then known as Maryland State College, in 1960 after having taught in the public schools of Ponca City, Oklahoma, and Oklahoma State University. Since coming to UMES, he has served as an instructor in mathematics, chairman of the Department of Mathematics, dean of student affairs, and chairman of the Division of Liberal Studies. Hytche was recently appointed by President Bush to serve on his Board of Advisers on Historically Black Colleges and Universities. He is currently chair of the Mideastern Athletic Conference Council of Presidents and Chancellors. He provided leadership for the 1890 Universities when he was chair of the Council of 1890 Presidents and Chancellors from 1985 to 1990.

ARTHUR KELMAN is a university distinguished scholar in plant pathology at North Carolina State University. He received a B.S. degree in biology from the University of Rhode Island and M.S. and Ph.D. degrees in plant pathology from North Carolina State University. In 1949, Kelman joined the faculty at North Carolina State University in the Department of Plant Pathology, where he rose from assistant professor to professor and was Reynolds Distinguished Professor from 1961 to 1965. In 1965 he was appointed professor and chairman of the Department of Plant Pathology at the University of Wisconsin-Madison. He was named L. R. Jones Distinguished Professor in 1975 and Wisconsin Alumni Research Foundation Senior Research Professor in 1984. Kelman returned to North Carolina State University in 1989 to assume his current position in the Department of Plant Pathology. Kelman has held leadership roles in numerous societies, and currently is chairman of Class VI: Applied Biological and Agricultural Sciences, National Academy of Sciences. Kelman's honors include membership to the National Academy of Sciences; the Outstanding Instructor Award and Distinguished Classroom Teacher Award, North Carolina State University; and the Spitzer Excellence in Teaching Award and the College of Agricultural and Life Sciences Amoco Distinguished Teaching Award, University of Wisconsin-Madison.

HARRY O. KUNKEL is dean emeritus and professor of life sciences in the College of Agriculture and Life Sciences, Texas A&M University. For over two decades he served as associate director and director of the Texas Agricultural Experiment Station and dean of agriculture. Currently in the departments of Animal Science and of Biochemistry and Biophysics, he teaches the undergraduate courses Principles of Animal Nutrition and Food and Humanity, the latter course being open to any major in the university. He also teaches a graduate course on contemporary issues in animal agriculture. Educated as a biochemist with a Ph.D. from Cornell University, Kunkel has recently written on human values in agricultural research and in setting nutritional policy. He served as senior education consultant to the U.S. Department of Agriculture's Project Interact.

JOSEPH E. KUNSMAN, JR., is associate dean for resident instruction, College of Agriculture, University of Wyoming. He received a B.S. degree from the Pennsylvania State University and M.S. and Ph.D. degrees from the University of Maryland. He joined the College of Agriculture at the University of Wyoming as assistant professor of food science in 1966, was promoted to professor in 1976, and was named associate dean for resident instruction in 1981. Since then he has also served 2 years as acting head of the Department of Home Economics. Kunsman is currently chair of the Resident Instruction Committee on Organization and Policy, Division of Agriculture. He served 6 years on the U.S. Department of Agriculture's (USDA) Joint Council National Higher Education Committee and 5 years on the USDA Joint Council on Food and Agricultural Sciences,

267

including 4 years as a member of the Executive Committee. Kunsman has served on numerous university committees including the Freshman Orientation Review Committee, the Honors/Scholars Committee, the Recruitment Task Force, and the University Coalition for Academic Success, where he chaired the Freshman Year Subcommittee. Kunsman has received several honors, including the John Ellbogen Outstanding Classroom Teaching Award and the Division of Student Affairs Outstanding Service Award.

JANIS W. LARIVIERE earned a B.S. degree from the University of Iowa and an M.S. degree from Drexel University. She is a biology teacher at Anderson High School in Austin, Texas, and 1991 recipient of the Anderson High School Teacher of the Year Award. A high school science teacher for the past 20 years, Lariviere has also been a teaching assistant in microbiology at Drexel University and a research assistant in cancer research at the University of Iowa and Thomas Jefferson University. During the past 15 years, she has presented 16 workshops to teachers and students. Her experience with developing textbooks and curricula is extensive. Lariviere has received numerous awards and honors, including the Tracor Scholar Award for Teaching Excellence, Travis County Engineering Society Outstanding Science Teacher Award, National Science Teachers Award for Innovations in Science Teaching, Texas Outstanding Biology Teacher Award, state finalist in the Presidential Awards for Excellence in Science Teaching, Tandy Technology Scholar, and a GTE GIFT program grant for implementation of innovative ideas in math and science teaching.

JAMES G. LEISING is supervisor of teacher education in agriculture at the University of California, Davis, and has a distinguished record of teaching in agricultural education at both the undergraduate and graduate levels. Currently, Leising provides leadership for four major research and development projects focused on the improvement of secondary and community college agricultural curricula. He has published in professional journals and developed curriculum materials for secondary and community college teachers. Leising has also presented numerous workshops to agricultural teachers and presented papers at regional and national professional meetings. He has been active in professional organizations, serving as secretary of the American Association of Teacher Educators in Agriculture, as president of its western region, and as a member of its board of directors. Leising received a B.S. degree in agricultural education from the University of Nebraska at Lincoln, taught secondary vocational agriculture in Nebraska, and completed M.S. and Ph.D. degrees in agricultural education and adult education at Iowa State University.

ROBERT J. MATTHEWS received a Ph.D. degree from Cornell University in 1974. He is professor of philosophy and member of the graduate faculties of philosophy and psychology at Rutgers University. He holds graduate degrees in engineering, French literature, and philosophy and has held visiting appointments at several universities, including Harvard, Massachusetts Institute of Technology, University of Western Ontario, and University of Bielefeld in Germany. Matthews' research activities are focused in three areas: the foundations of cognitive science, theoretical psycholinguistics, and ethical issues in agricultural and environmental policy. In the latter area of research, he is particularly interested in the way in which ethical considerations can and should influence the development and implementation of public policy and the ethical acceptability of certain economic criteria for choosing among available policy options. He is the author of numerous scholarly papers, editor of *Learnability and Linguistic Theory* (Norwell, Mass: Kluwer Academic, 1989), and coauthor of *Public Policy, Ethics, and Agriculture* (New York: Macmillan, in press).

EDNA L. McBREEN is director of the Office of International Programs at West Virginia University. Prior to coming to West Virginia in 1988, McBreen had extensive professional experience in domestic and international education with a focus on agriculture, home economics, and adult education; training; and extension. She had worked as senior associate with Creative Associates in Washington, D.C., where she developed cooperative linkages among universities, colleges of agriculture and home economics, and agribusiness organizations. McBreen also was an agricultural education specialist for the Africa Bureau of the U.S. Agency for International Development, assistant professor in the Department of Agricultural Education at Texas A&M University, and instructor in the Department of Home Economics at Southwest Texas State University. McBreen received a B.S. degree in home economics education from Cornell University, a M.Ed. degree in adult and extension education from Texas A&M University, and a Ph.D. degree in human service studies from Cornell University. She is on the board of directors of the Association for International Agricultural and Extension Education and a member of the American Association of Teacher Educators in Agriculture.

RICHARD H. MERRITT is professor of horticulture, Cook College, Rutgers University. His academic career has spanned 33 years, most of it at Rutgers University, with visiting professorships and visiting academic administrator roles at the University of California, Davis, and at the University of Puerto Rico. He has held full- and part-time academic leadership roles for much of that time. For 20 years, he was director of resident instruction for the College of Agriculture and Environmental Science and dean of instruction, Cook College, Rutgers University. Merritt's part-time administrative roles at the national level have included director of the National Agriculture and Natural Resources Curriculum Project and executive liaison consultant for the U.S. Department of Agriculture's Project Interact, a national curriculum revitalization program. At the international level, he was alternative chair of the Title XII U.S. Agency for International Development Joint Committee on Agriculture Development and chair of teams that designed research and teaching programs for colleges overseas. Throughout his career, Merritt has written and lectured about research on ornamental horticultural plants and academic innovation and program revitalization in agriculture, natural resources, environmental studies, and the life sciences. In 1989 he was commissioned by the Office of Technology Assessment, U.S. Congress, to write a paper on integrating agricultural and environmental studies in colleges of agriculture and natural resources.

PEGGY S. MESZAROS is dean of the College of Human Environmental Sciences at the University of Kentucky. She received a B.S. degree from Austin Peay State University, an M.S. degree from the University of Kentucky, and a Ph.D. degree from the University of Maryland. She has pursued a teaching and administrative career in Kentucky, Pennsylvania, Germany, Nebraska, Maryland, and Oklahoma. Her accomplishments at Kentucky include authoring over 50 journal articles, two books, and numerous national research presentations and major leadership roles in the American Home Economics Association, the Association of Administrators of Home Economics, the advisory board of the National Association of Extension Home Economists, the National Higher Education Committee, and the National Extension Committee. She is currently vice-president for public affairs of the American Home Economics Association and serves on the Executive Committee of Project 2000, a minority recruitment effort for the field of home economics.

GARY E. MILLER is associate vice-president for instructional development at the University of Maryland University College. He serves as executive director of the International University Consortium, a 40-member course development and delivery consortium, and administers the development of open learning courses and technology-based education materials for University College. He also heads the Institute for Distance Education, which coordinates distance education activities among the 11 institutions of the University of Maryland system. Miller is the author of *The Meaning of General Education: The Emergence of a Curriculum Paradigm* (New York: Teachers College Press, 1988) and numerous articles and papers on undergraduate curriculum and distance education issues. He earned a doctor of education degree in higher education from the Pennsylvania State University.

WILLIAM H. MOBLEY is the twentieth president of Texas A&M University. He earned the B.A. degree in psychology and economics from Denison University and master's and Ph.D. degrees in industrial-organizational psychology from the University of Maryland. Mobley served as corporate manager of employee relations research and planning for PPG Industries. From 1973 to 1980, he served as director of the Center for Management and Organizational Research at the University of South Carolina. He came to Texas A&M University in 1980 as head of the Department of Management, and in 1983 he became dean of the College of Business Administration. Mobley served as executive deputy chancellor of the Texas A&M University system from 1986 to 1988 and was named president of Texas A&M University in August 1988. His research and writing on organizational behavior and effectiveness are cited frequently. He is also vice-chairman of the Texas Association of University Chancellors and Presidents, vice-president/president-elect of the Association of Texas Colleges and Universities, and a member of the Council of President's Executive Committee of University Research Associates, Inc. In 1990, President Bush appointed Mobley to a 2-year term on the U.S. Commission on Minority Business Development.

LAURENCE D. MOORE is professor of plant pathology and head of the Department of Plant Pathology, Physiology, and Weed Science at Virginia Polytechnic Institute and State University. He graduated from the University of Illinois with a B.S. degree in horticulture and continued his education at Pennsylvania State University, where he received M.S. and Ph.D. degrees. He has been at Virginia Tech since 1965 and became the department head in 1985. He has taught undergraduate and graduate courses in stress physiology, plant metabolism, and disease physiology. Moore's research interests include air pollution, fungal diseases, and disease physiology.

JAMES R. MOSELEY was appointed assistant secretary of agriculture for natural resources and environment by President George Bush on July 2, 1990. As assistant secretary, Moseley is responsible for directing the policies and supervising the activities and programs of the U.S. Forest Service and the U.S. Soil Conservation Service. Before appointment to his present position, Moseley served as agricultural adviser to William Reilly, administrator of the U.S. Environmental Protection Agency, a position in which he advised the administrator on environmental issues that directly affected the agricultural industry. Moseley graduated from Purdue University in 1973 with a B.S. degree in horticulture. Following graduation, Moseley began a farming operation in Indiana that today is a grain and hog enterprise. He has also served with several public policy groups that work on agriculture and rural development policy at the local, state, and national levels.

C. JERRY NELSON is curator's professor in the Department of Agronomy, University of Missouri. He earned a B.S. degree in animal husbandry and an M.S. degree in forage production from the University of Minnesota. His Ph.D. degree in forage physiology was awarded by the University of Wisconsin-Madison in 1966. Nelson then joined the faculty at the University of Missouri in 1967, advancing to full professor in 1975 and curator's professor in 1989. Nelson was a visiting research scientist at the Welsh Plant Breeding Station, Aberystwyth, Wales, from 1973 to 1974 and academic guest at the Swiss Institute of Technology, Zurich, from 1980 to 1981.

MORT H. NEUFVILLE has been dean and 1890 research director of the School of Agricultural Sciences, University of Maryland, Eastern Shore, since 1983. Born in Portland, Jamaica, West Indies, he attended the Jamaica School of Agriculture, where he obtained a diploma in agriculture. He received a B.S. degree from Tuskegee Institute and M.S. and Ph.D. degrees in animal science from the University of Florida. Neufville served as assistant professor of animal science and head of animal science at Prairie View A&M University from 1974 to 1978 and then as associate dean of applied science and technology at Lincoln University in Missouri from 1978 to 1983. He is also project manager of the Cameroon Root and Tuber Food Crops Research Project and associate director of the Maryland Agricultural Experiment Station. Neufville is a chairman and member of the Association of 1890 Deans of Agriculture/ 1890 Research Directors. He serves on many other national committees, including the Northeast Regional Council and the National Higher Education Committee, which are subcommittees of the Joint Council on Food and Agriculture.

DIANA G. OBLINGER is an academic discipline specialist in agriculture and life sciences at the Institute for Academic Technology. Her areas of expertise are biology, research and teaching in agriculture, and veterinary medicine. She received her academic training at Iowa State University, earning a B.S. degree in biology, an M.S. degree in plant breeding, and a Ph.D. degree in plant breeding and cytogenetics. Her previous professional positions include associate dean and director of resident instruction as well as associate professor of agronomy at the University of Missouri-Columbia, associate professor of horticulture at Michigan State University, plant breeder at DeKalb AgResearch, Inc., and adjunct professor at North Carolina State University and at Clemson University. A member of the American Society of Agronomy and the National Association of Colleges and Teachers of Agriculture, Oblinger has received several academic awards, including the Burlington Northern Outstanding Teacher Award, E. F. Cooper Academic Innovation Award, and Gamma Sigma Delta Outstanding Young Researcher Award.

FRANK PRESS, president of the National Academy of Sciences in Washington, D.C., has advised four presidents on scientific issues and has made pioneering contributions in several fields. He has been named most influential American scientist in annual surveys by *U.S. News and World Report* three times, most recently in 1985. Press is recognized internationally for his study of the sea floor and the earth's crust and deep interior, and has made contributions in geophysics, oceanography, lunar and planetary sciences, and natural resource exploration. In 1977, Press was appointed President Carter's science adviser and director of the Office of Science and Technology Policy. He served on science advisory committees during the Kennedy and Ford administrations and was appointed by President Nixon to the National Science Board, the policymaking body of the National Science Foundation. Press participated in bilateral science agreement negotiations with China and the Soviet Union and was a member of the U.S. delegation to the nuclear test ban negotiations in Geneva and Moscow. He graduated from City College of New York with a degree in physics and received advanced degrees from Columbia University. Press joined Columbia's faculty in 1952 and 3 years later was appointed professor of geophysics at the California Institute of Technology. He joined the Massachusetts Institute of Technology in 1965 and in 1982 was named institute professor, a title reserved for scholars of special distinction. Press is coauthor of the textbook *Earth* (New York: W. H. Freeman, 1985), which is widely used in American and foreign universities.

JAMES L. RAINEY, former president and chief executive officer of Farmland Industries, the Kansas City-based farm supply and marketing cooperative, was born in Nashville, Tennessee, and grew up on a small Indiana farm near Indianapolis. He received a B.S. degree in agriculture from Purdue University in 1952. Rainey began his professional career in 1954 as a sales representative with Allied Chemical Corporation. He has held sales and management positions in the agrichemical industry throughout his career and was president of Kerr-McGee Chemical and senior vice-president of Kerr-McGee Corporation before joining Farmland in 1986. Rainey has been an active volunteer in civic affairs throughout his career, with special interest in education and health care initiatives. He

273

recently retired from Farmland, but he continues working in voluntary service and business consulting.

SUSAN G. SCHRAM is assistant director of federal relations for international affairs at the National Association of State Universities and Land-Grant Colleges. She holds bachelor's and master's degrees from Michigan State University and a Ph.D. degree from the University of Maryland and has 15 years' experience at the county, state, and national levels. Schram has served as a county extension agent and program leader in the Michigan Cooperative Extension Service and came to Washington, D.C., in 1980 to serve as staff to the Joint Council on Food and Agricultural Sciences. She has worked as a consultant in the Washington, D.C., area since 1982 and, most recently, at the University of Maryland as special assistant to the vice-chancellor while completing her Ph.D program.

NORMAN R. SCOTT is vice-president for research and advanced studies at Cornell University. He served as director of the Cornell University Agricultural Experiment Station and director of the Office for Research, College of Agriculture and Life Sciences, from 1984 to 1989 after having served nearly 7 years as chairman of the Department of Agricultural and Biological Engineering. He received a B.S.A.E. degree with honors from Washington State University in 1958 and a Ph.D. degree from Cornell University in 1962. He has been a member of the Cornell faculty since 1962 in the Department of Agricultural and Biological Engineering. Sabbatical leaves have been spent in the Department of Biomedical Engineering at Case Western Reserve University and at the National Institute for Research in Dairying, Reading, England. Scott has been involved in bioengineering research for over 20 years. Recent projects have included electronic applications in agriculture. Scott was elected technical vice-president of the American Society of Agricultural Engineers in 1989 for a 3-year term. He received the Henry Giese Award in 1989, and was elected to the National Academy of Engineering in 1990.

SAMUEL H. SMITH is president of Washington State University. Previously, he was dean of the College of Agriculture, director of the Agricultural Experiment Station, and director of the Cooperative Extension Service at the Pennsylvania State University. Smith has also been professor and head of the Department of Plant Pathology at the Pennsylvania State University and assistant professor of plant pathology at the University of California, Berkeley, where he had received B.S. and Ph.D. degrees in plant pathology. In 1965 he was a postdoctoral fellow at the North Atlantic Treaty Organization, Sussex, England, and in 1989 he received an honorary doctoral degree from Nihon University, Tokyo. Smith serves on several committees of the National Association of State Universities and Land-Grant Colleges. He is on the board of directors of the Washington China Relations Council, the Economic Development Partnership for Washington, and the Washington International Ag-Trade Center, and is a member of several committees, including the Steering Committee for Regional Telecommunication Cooperative of the Western Interstate Commission for Higher Education, the American Council on Education's Commission on Women in Higher Education, the Washington State International Trade Assistance Advisory Committee, and the Council on Competitiveness.

PETER SPOTTS is national news editor for *The Christian Science Monitor*. Since he joined the *Monitor* in 1976, Spotts has served in various jobs: as a Midwest correspondent; as staff editor in national news, specializing in science and technology, defense and arms control, and economics; and as special projects editor. In February 1987 he accepted a newly created science writing post. The following November, Spotts and three *Monitor* colleagues received the Forum Award of the U.S. Council for Energy Awareness for an April 1987 series exploring the future of nuclear energy after Chernobyl. Spotts was an editorial writer prior to becoming national news editor.

PAUL B. THOMPSON is director of the Center for Biotechnology Policy and Ethics and associate professor of philosophy and agricultural economics at Texas A&M University. He earned a B.A. degree in philosophy from Emory University and M.A. and Ph.D. degrees in philosophy from the State University of New York at Stony Brook. Prior to joining the Texas A&M faculty in 1982, Thompson was visiting assistant professor of philosophy at Texas A&M, and in 1986 he was visiting scholar at the U.S. Agency for International Development. Thompson's honors include president of the Food, Agriculture, and Human Values Society; participant in the summer seminars of the National Endowment for the Humanities; fellow of the Council on Foreign Relations and International Affairs; and resident fellow of the National Center for Food and Agricultural Policy, Resources for the Future.

RAY THORNTON was elected to the U.S. House of Representatives in 1990 from Arkansas' Second District. He is not a newcomer to the U.S. Congress, having served three terms in the House from 1972 to 1979. In 1978 he ran unsuccessfully for the U.S. Senate. Thornton began his public service career as a deputy prosecutor for Pulaski and Perry counties in 1956 and was elected Arkansas attorney general in 1970. His congressional experience during the 1970s included chairing the House Subcommittee on Science, Research, and Technology, and serving on the House Agriculture and

Judiciary committees. The 1980s became Thornton's decade as an educator. From 1979 to 1980, he directed the Ouachita Baptist University and Henderson State University Joint Educational Consortium, was president of Arkansas State University from 1980 to 1984, and served as president of the University of Arkansas system from 1984 to 1989. Science-related achievements during the 1980s included chairing the Committee on Science, Engineering, and Public Policy of the American Association for the Advancement of Science. His congressional assignments today include the House Science, Space, and Technology Committee, Government Operations Committee, and Democratic Steering and Policy Committee. Thornton has a degree in international relations from Yale University and a law degree from the University of Arkansas.

ANNE M. K. VIDAVER is professor and head of the Department of Plant Pathology, University of Nebraska-Lincoln. She served as the interim director of the Center for Biotechnology from 1988 to 1990. A native of Vienna, Austria, Vidaver graduated from Russell Sage College with a B.A. degree in biology, followed by M.A. and Ph.D. degrees in bacteriology with minors in plant physiology from Indiana University-Bloomington. Prior to joining the faculty at the University of Nebraska-Lincoln, Vidaver worked at Brookhaven National Laboratory and as a research associate in plant pathology at the University of Nebraska. She is president of the American Phytopathological Society and a member of the Board on Agriculture of the National Research Council. Her research interests have focused principally on plant-associated bacteria. She is adviser or consultant to several companies and federal agencies and is a member of the National Institutes of Health's Recombinant DNA Advisory Committee and the U.S. Department of Agriculture's Agricultural Biotechnology Research Advisory Committee. Vidaver has authored or coauthored over 130 scientific articles and one book.

DONALD M. VIETOR is a crop physiologist in the Soil & Crop Sciences Department of Texas A&M University. He was a member of the Systems Task Force of the National Agriculture and National Resources Curriculum Project from 1982 to 1986 and a contributor to the recent book *Systems Approaches for Improvement in Agriculture and Resource Management* (New York: Macmillan, 1990). He is currently participating in a research project, Ethics in Agriculture: Holistic and Experiential Approaches, under the sponsorship of the office of Higher Education Programs of the U.S. Department of Agriculture.

CONRAD J. WEISER is dean of the College of Agricultural Sciences at Oregon State University. Until his appointment as dean in 1991, he was professor and head of the Department of Horticulture, Oregon State University. Previously, he was professor in the Department of Horticulture and Laboratory of Plant Hardiness at the University of Minnesota, where he established the Laboratory of Plant Hardiness, taught graduate and undergraduate courses, and consulted and traveled internationally on research and graduate program planning. In his current position at Oregon State, Weiser provides liaison with Oregon commodity commissions and other producer and processor organizations, and he established the 70-member Industry Advisory Board that represents horticultural producers, processors, suppliers, and consumers in Oregon. He is adviser to the National Science Foundation's International Programs Division and external reviewer of research and educational programs at universities in 11 U.S. states and Canadian provinces and at the International Potato Research Center in Peru. He is a member of the Board on Agriculture of the National Research Council. Weiser received a B.S. degree from North Dakota State University in horticulture and a Ph.D. degree from Oregon State University in plant pathology and physiology.

PAUL H. WILLIAMS is professor of plant pathology at the University of Wisconsin-Madison and serves as the director of the Center for Biology Education (CBE). Established in 1989, the CBE's mission is to improve teacher and student education in the biological sciences. Williams developed the rapid-cycling brassicas (RCBs) that have up to 10 reproductive cycles per year and serve as models for his research on the genetics of plant-parasite interactions. In 1983 he established the Crucifer Genetics Cooperative, a network that distributes RCB seed and information to over 1,600 researchers in 48 countries. Using the RCBs, he initiated the Wisconsin Fast Plants and Bottle Biology programs in 1988, both supported by the National Science Foundation. These programs aim to increase the involvement of students in the biological sciences. Williams has authored or coauthored over 165 publications and 13 books. He teaches three courses at the University of Wisconsin-Madison and received a Distinguished Teaching Award there in 1990. Williams received a B.S.A. degree in plant science from the University of British Columbia and a Ph.D. degree in plant pathology from the University of Wisconsin-Madison. He did postdoctoral work in plant biochemistry at the Boyce Thompson Institute for Plant Research in Yonkers, New York.

EDWARD M. WILSON is deputy administrator for the Cooperative State Research Service, U.S. Department of Agriculture. His responsibilities include providing agency-wide leadership for the plant and animal sciences programs. Wilson earned B.S. and M.S. degrees from McGill University in Canada in 1964 and 1966, respectively, majoring in animal science. He earned a Ph.D. degree in

277

dairy science from Ohio State University in 1969. Wilson began his career with the Agriculture Department at Tuskegee University, where he was assistant professor and director of the Guyana Ranch Management Program. He later became dean of the College of Applied Sciences and Technology at Lincoln University in Missouri. At Lincoln he also was dean of cooperative extension, agricultural research, and international programs. Wilson has extensive experience in international development and has traveled widely as an adviser and consultant on agricultural matters. As a member of a Tuskegee University team, he visited the Republic of South Africa in 1974 and studied public education and agricultural development in the homelands (areas designated for blacks). In 1984, Wilson served as senior scientist for agricultural studies conducted in eight African countries. He has authored numerous papers and scientific publications in agriculture and related fields.

ALVIN L. YOUNG was detailed from the Executive Office of the President to the Office of the Secretary of Agriculture in November 1987 as the director of the Office of Agricultural Biotechnology. He assumed the permanent post of scientific director and science adviser on June 1, 1989. In addition to directing the activities of the office, Young serves as executive secretary of the Agricultural Biotechnology Research Advisory Committee, U.S. Department of Agriculture (USDA), and chairman of USDA's Biotechnology Council. He received a B.S. degree in agricultural sciences and an M.S. degree in agronomy from the University of Wyoming. He then earned a Ph.D. degree in herbicide physiology from Kansas State University. Young has conducted extensive research on the environmental, toxicological, and human health effects of insecticides and herbicides and was director of research for environmental issues at the Veterans Administration. He was the senior policy analyst for life sciences with the White House Office of Science and Technology Policy, and he has served as consultant or adviser to six federal agencies and the National Research Council. Young has authored many scientific books and articles on environmental issues, risk assessment, and science policy.

B

Poster Exhibits

Overview of Project Sunrise
C. Eugene Allen and Richard Simmons, *University of Minnesota*

Adult Learners in the College Classroom and Coping with Academic Stress
F. H. Buelow, *University of Wisconsin*

Enroll in Internship Program
James Diamond, *Pennsylvania University*

Field Study in Tropical Agriculture
Arlen Etling, *Pennsylvania State University*

Integration for Humanity
Gladys Gonzalez, *University of Puerto Rico*

Innovative Course Ethics in Agriculture
W. J. Hanekamp, *University of Arizona*

Bioresources Research Program
John B. Hayes, *Oregon State University*

Minorities in Agriculture and Natural Resources Association
William L. Henson, *Pennsylvania State University*

Be a Master Student
Marianne Houser and Barbara Wade, *Pennsylvania State University*

Pennsylvania Governor's School for Agricultural Sciences
Marianne Howard and Tracey Hoover, *Pennsylvania State University*

Science in Agriculture Symposium
Paul Hummer, *Oklahoma State University*

The Curriculum Responds to Society
Taylor Johnston and Susan De Rosa, *Michigan State University*

Diversity and Pluralism
Taylor Johnston and Susan De Rosa, *Michigan State University*

International Study Programs
Taylor Johnston and Susan De Rosa, *Michigan State University*

Horizons Unlimited
Paul R. Kohn, *University of Arizona*

The Advising Specialist
Cheryl Kolbe, *Oregon State Unversity*

Globalization of a Curriculum
James McKenna, *Virginia Polytechnic Institute*

Teaching by Satellite
L. H. Newcomb, *Ohio State University*

A Holistic Model for Training Minorities in Agribusiness
Zaach Olorunnipa, *Florida A&M University*

A New Major in Environmental Science
Davis Parrish, *Virginia Polytechnic Institute*

Kharkov Institute Exchange
Leroy Rogers, *Washington State University*

A Case Study "Capstone" Course in Crop Management
Steve Simmons, *University of Minnesota*

Landscape Contracting
Dan Stearns, *Pennsylvania State University*

Computer-Aided Learning Modules Assist in Plant Identification
T. Davis Sydnor, Scott Biggs, and L. H. Newcomb, *Ohio State University*

Mid-America International Agriculture Consortium Agricultural Travel Course
Steven J. Thien, *Kansas State University*

Strengthening College of Tropical Agriculture and Human Resources Undergraduate Program
Sylvia Yuen and Dian Nafis, *University of Hawaii*